ANNALS *of* THE NEW YORK ACADEMY OF SCIENCES

T0310253

EDITOR-IN-CHIEF
Douglas Braaten

ASSOCIATE EDITOR
Rebecca E. Cooney

PROJECT MANAGER
Steven E. Bohall

EDITORIAL ADMINISTRATOR
Daniel J. Becker

Artwork and design by Ash Ayman Shairzay

The New York Academy of Sciences
7 World Trade Center
250 Greenwich Street, 40th Floor
New York, NY 10007-2157

annals@nyas.org
www.nyas.org/annals

The New York Academy of Sciences

Published by Blackwell Publishing
On behalf of the New York Academy of Sciences

Boston, Massachusetts
2012

ANNALS *of* THE NEW YORK ACADEMY OF SCIENCES

VOLUME
1256

ISSUE

The Year in Evolutionary Biology

ISSUE EDITORS
Timothy A. Mousseau[a] and Charles W. Fox[b]

[a]University of South Carolina and [b]University of Kentucky

TABLE OF CONTENTS

Academy Membership: Connecting you to the nexus of scientific innovation

Since 1817, the Academy has carried out its mission to bring together extraordinary people working at the frontiers of discovery. Members gain recognition by joining a thriving community of over 25,000 scientists. Academy members also access unique member benefits.

 Network and exchange ideas with the leaders of academia and industry

 Broaden your knowledge across many disciplines

 Gain access to exclusive online content

 Select one free *Annals* volume each year of membership and get additional volumes for just $25

Join or renew today at **www.nyas.org**.
Or by phone at **800.843.6927** (**212.298.8640** if outside the US).

Ann. N.Y. Acad. Sci. ISSN 0077-8923

ANNALS OF THE NEW YORK ACADEMY OF SCIENCES
Issue: *The Year in Evolutionary Biology*

Genome and gene duplications and gene expression divergence: a view from plants

Yupeng Wang,[1,2] Xiyin Wang,[1,3] and Andrew H. Paterson[1,2,4]

[1]Plant Genome Mapping Laboratory and [2]Institute of Bioinformatics, University of Georgia, Athens, Georgia. [3]Center for Genomics and Computational Biology, School of Life Sciences and School of Sciences, Hebei United University, Tangshan, Hebei, China. [4]Department of Plant Biology, Department of Crop and Soil Sciences, Department of Genetics, University of Georgia, Athens, Georgia

Address for correspondence: Andrew H. Paterson, Plant Genome Mapping Laboratory, University of Georgia, 111 Riverbend Road, Athens, Georgia 30602. paterson@uga.edu

With many plant genomes sequenced, it is now clear that one distinguishing feature of angiosperm (flowering plant) genomes is their high frequency of whole-genome duplication. Single-gene duplication is also widespread in angiosperm genomes. Following various mechanisms of gene duplication, genes are often retained or lost in a biased manner, which has suggested recent models for gene family evolution, such as functional buffering and the gene balance hypothesis in addition to now-classical models, including neofunctionalization and subfunctionalization. Evolutionary consequences of gene duplication, often studied through analyzing gene expression divergence, have enhanced understanding of the biological significance of different mechanisms of gene duplication.

Keywords: whole-genome duplication; single-gene duplication; plant; gene expression; divergence; gene retention; evolutionary model

Introduction

Whole-genome duplications (WGDs) have occurred in the lineages of plants,[1] animals,[2,3] and fungi.[4,5] Major consequences of WGDs can include evolution of novel or modified gene functions[6–9] and/or provision of buffer capacity[10,11] or genetic redundancy[12–17] that increases genetic robustness. WGDs may also increase opportunities for non-reciprocal recombination,[18] permitting or causing duplicated genes to evolve in concert for a period of time.[19,20] Rapid DNA loss and restructuring of low-copy DNA,[21–24] retrotransposon activation,[25–27] and epigenetic changes[28–33] following WGD may further provide materials for evolutionary change. In addition to WGDs, genes may be duplicated by other mechanisms, which have been collectively referred to as small-scale duplications or single-gene duplications.[34–36]

The angiosperms are an outstanding model in which to elucidate the consequences of genome and gene duplications. All angiosperms have experienced paleopolyploidy (i.e., ancient WGDs that occurred at least several million years ago),[37] and many angiosperm genomes have experienced multiple WGDs.[38,39] For example, *Arabidopsis*, selected as the first angiosperm genome to be sequenced due in part to its small genome size and minimal DNA sequence duplication, has experienced two "recent" WGDs, that is, since its divergence from other members of the Brassicales clade (α and β), and a more ancient triplication (γ) shared with most if not all eudicots.[37,39,40] Likewise, rice appears to have experienced at least two WGDs, one shared with most if not all cereals (ρ) and another more ancient event (σ).[41] A recent study provided evidence for two additional ancient WGDs for angiosperms and seed plants, respectively.[42] Single-gene duplications are also widespread in angiosperms.[36,43]

In eukaryotes, changes in gene expression directed by transcriptional regulation often give rise to new phenotypes, with gene expression divergence often used as a proxy indicator of the divergence of gene functions. Thus, one avenue for systematic investigation of the consequences of genome and gene duplications can be comparison of spatiotemporal

doi: 10.1111/j.1749-6632.2011.06384.x
Ann. N.Y. Acad. Sci. 1256 (2012) 1–14 © 2012 New York Academy of Sciences.

expression patterns between duplicated genes and subsequently relating gene expression divergence to underlying genetic mechanisms and evolutionary models.

Here, we review recent advances in the understanding of gene duplication mechanisms and their different evolutionary consequences as revealed by analysis of gene expression divergence. First, we introduce various mechanisms underlying gene duplications. Then, we review evolutionary models for duplicate gene retention. Last, we review recent advances in understanding expression divergence between duplicate genes associated with various genetic mechanisms.

Gene duplication modes

Compared to other taxa, a distinctive feature of angiosperms is the relatively high frequency at which WGDs have been preserved during their evolution.[44] It has been suggested that all angiosperms are paleopolyploids[37] and that WGD may be a cyclic process in these taxa.[10] However, computational identification of WGD events, as well as the duplicate genes that were created and retained from WGDs, has been a challenging task.[45] In general, this task is often solved through analyzing synteny (i.e., genes remaining on corresponding chromosomal regions) and collinearity (i.e., genes remaining in corresponding orders along the chromosomes) among several related species.[39,40] One classical method for synteny detection is to use all versus all BLASTP searches as inputs and model the matches in a homology matrix for synteny detection through clustering neighboring matches inside the matrix. This approach was implemented in ADHoRe,[46] DiagHunter,[47] and other derived algorithms.[48] Another classical method for synteny detection is to use dynamic programming to detect synteny and statistical strategies to evaluate synteny—for example, DAGchainer[49] and ColinearScan.[50] However, the aforementioned tools detect only pair-wise collinear segments, and thus are insufficient for distinguishing different WGD events that a genome experienced.

Early approaches for computational detection of paleopolyploidy were "bottom-up," starting with the most recent duplication event, and then resolving more ancient ones sequentially through recursively merging duplicated segments to generate hypothetical intermediate chromosomal segments

need be generated.[3,37] Alternatively, top-down algorithms, available in the MCscan software,[40] can be used instead.[39,40] Pair-wise collinear segments are picked from whole-genome BLASTP results through dynamic chaining of collinear gene pairs[49] to produce multialignments of collinear segments by fixing one gene order as reference and then heuristically stacking the pair-wise collinear segments one after another. The reference gene order, which is used to thread the multialignments of collinear segments, should be relatively conserved, often available from a related species (outgroup) that is assumed to have experienced less-intensive WGDs and chromosome rearrangements. An advantage of MCScan is that it can reveal cryptic synteny based on transitive homology, which has been referred to as "ghost duplications."[46,51,52] MCScan was first used to analyze the duplication relationships among *A. thaliana*, *Populus trichocarpa*, and *Carica papaya*, using *Vitis vinifera* as the reference gene order.[40] A shared ancient hexaploidy (γ) event was revealed among these taxa. The implementation of MCscan also revealed that proportions of genes created and retained from WGDs or segmental duplications fluctuated among taxa. For example, 54% of *Arabidopsis* genes and 80% of *Populus* genes were created by WGDs or segmental duplications, versus only 11% of *Carica* genes and 18% of *Vitis* genes. Segmental duplication can be regarded as a type of small-scale duplications, but which is often difficult to distinguish from WGD in angiosperms. Note that the idea of using an outgroup to facilitate the identification of WGD segments has also been implemented in yeast.[5,53]

In addition to WGD and segmental duplication, other gene duplication modes are widely referred to as single-gene duplications.[35,36] Tandem duplication generates gene copies that are consecutive in the genome and is presumed to arise through unequal chromosomal crossing over.[36] Tandem duplicates may account for ~10% of *Arabidopsis* or rice genes[54] and may contribute to the expansion of some large gene families.[55] Dispersed duplicates are neither adjacent to each other in the genome nor within homeologous chromosome segments.[56] Distant single-gene transposition may explain part of the widespread existence of dispersed duplicates within and among genomes.[36,43] Such distantly transposed duplications may occur by DNA-based or RNA-based mechanisms.[35] DNA transposons such as packmules (rice),[57] helitrons (maize),[58] and CACTA elements (sorghum)[27] may relocate duplicated genes or gene segments to new chromosomal positions (i.e., DNA-based transposed duplication). RNA-based transposed duplication, often referred to as retrotransposition, typically creates a single-exon retrocopy from a multiexon parental gene by reverse transcription of a spliced messenger RNA. It is presumed that the retrocopy duplicates only the transcribed sequence of the parental gene, detached from the parental promoter. The new retrogene is often deposited in a novel chromosomal environment with new (i.e., nonancestral) neighboring genes and, having lost its native promoter, is only likely to survive as a functional gene if a new promoter is acquired.[59,60]

However, to identify distantly transposed gene duplications at genomic scales is not an easy task. To accomplish this aim, genes at ancestral chromosomal positions first need to be discerned. The Plant Genome Duplication Database (PGDD, available at http://chibba.pgml.uga.edu/duplication) can be used for this task. A gene locus is regarded as ancestral if the resident gene along with any of its homologous genes (paralogs/orthologs) occurs at corresponding loci within any pair of collinear segments in PGDD. Using this criterion, one can discern genes at ancestral loci from those that are not and may have been transposed. For a pair of distantly transposed duplicate genes, one copy should be at its ancestral locus and the other should be at a nonancestral locus, named the parental copy and transposed copy, respectively. If the parental copy has more than two exons and the transposed copy is intronless, the pair of duplicate genes is likely to have occurred by retrotransposition (RNA-based transposition). If both copies have a single exon, the pair of duplicates cannot be classified into DNA-based or RNA-based transposed duplication. If both copies have multiple exons, the transposed duplication is likely to have occurred by DNA-based transposition.

We note that although many pairs of dispersed duplicates can be explained by distant single-gene transpositions, the underlying mechanisms for other dispersed duplicates are not well understood. For a pair of dispersed duplicates, both copies may be located at two different ancestral loci (i.e., collinear positions of two different within or across species collinear segments), neither of which is likely to have experienced transposition. One possibility is that they originated from very ancient WGDs that are no longer discernable based on collinearity approaches or from single-gene duplications that happened before speciation events. Another case is that both dispersed copies may be located at nonancestral loci. Though both seem to have experienced transposition, their ancestral copy cannot be identified in the genome to which they belong and may have been lost or altered beyond recognition.

Another gene duplication mode that could be distinguished is the case in which duplicate genes are near one another but separated by a few genes. This mode, lying between tandem and distantly transposed duplications, was referred to as proximal duplication.[61] In many studies, proximal duplication has been regarded as a form of tandem duplication. However, the underlying genetic mechanisms can be different between tandem and proximal duplications. Tandem duplication is presumed to arise through unequal chromosomal crossing over.[36] However, proximal duplicates are presumed to arise from ancient tandem duplicates interrupted by other genes[43] (similar to but indeed different from tandem duplication as interrupted genes may cause further genomic changes to ancient tandem duplicates) or localized transposon activities,[61] in contrast to distant transposon activities that can generate distantly transposed duplicates. We found that proximal duplicates have different features and evolutionary fates than tandem or dispersed duplicates.[62] Accordingly, we suggested that at least six gene duplication modes can be distinguished: WGD/segmental, tandem, proximal, DNA-based transposed, retrotransposed, and dispersed duplications.[62] Table 1 summarizes a brief comparison of different modes of gene duplication. Further, we found that the expansions of gene families may be enriched with particular modes of gene duplication.[62] For example, DNA-based transposed duplications are enriched in disease resistance gene homologs and the cytochrome P450 gene family (Fig. 1A–C), while WGD duplicates are enriched in the cytoplasmic ribosomal protein gene family and C2H2 zinc finger proteins (Fig. 1D–F). We then suggested that gene family members may have common nonrandom patterns of origin that may recur independently in different evolutionary lineages.[62]

Although it is clear that genes can be duplicated by various genetic mechanisms, currently the classification of gene duplication modes is mainly based

Table 1. Comparison of different modes of gene duplication

Mode of gene duplication	Feature	Mechanism	Change of promoter
Whole genome or segmental	Duplication of all genes in a genome or chromosomal region.	Abnormal cell division	No
Tandem	Duplication of a single gene. The newly created gene is adjacent to its parental gene.	Unequal chromosomal crossing over	No
Proximal	Duplication of a single gene. The newly created gene is near, but not adjacent to, its parental gene.	Tandem duplicates interrupted by other genes, or induced by localized transposon activities	Yes
DNA-based transposed	Duplication of a single gene. The newly created gene is translocated to a different and nonhomologous chromosomal region.	Induced by DNA transposons	Yes
Retrotransposed	Duplication of a single gene. The newly created gene is translocated to a different and nonhomologous chromosomal region.	Induced by retrotransposons	Yes
Dispersed	Duplication of a single gene. The newly created gene is translocated to a different and nonhomologous chromosomal region.	Unclear	Unclear

on measuring the physical distance between duplicate genes (the distance between collinear positions can be regarded as zero), which itself has limitations. First, a duplicate gene may be simultaneously located at a collinear position of a collinear segment and within a tandem array. Next, since transposons may relocate duplicate genes to distant chromosomal positions, it is reasonable to conjecture that transposons may also (perhaps are more likely to) relocate duplicate genes to adjacent or proximal chromosomal positions. So the mechanisms underlying tandem and proximal duplicates may be intermingled between unequal chromosomal crossing over and transposon activities. Further, a pair of duplicate genes may have experienced multiple genomic events. For example, two duplicates initially created by a WGD event might experience further gene movement or transposition. For such duplicates, a simple mode of gene duplication may be inappropriate. In addition, as mentioned above, the mechanisms underlying many dispersed duplicates are not well understood. These facts suggest that currently the associations between gene duplication modes and genetic mechanisms are generally rough, necessitating future efforts to depict a whole and clear picture of various modes and mechanisms of gene duplications.

Retention of duplicate genes

It is assumed that when a gene is duplicated, the two duplicates are almost identical in sequences and they have full redundancy (functional overlap). Since a likely consequence of gene duplication is reversion to single copy (singleton) status, the mechanisms for retention of duplicated genes have been a widely studied topic for evolutionary biologists. There are a few detailed reviews for models of duplicate gene retention.[63–66]

Population genetic theory suggests that full redundancy between duplicate genes is not evolutionarily stable.[67,68] Newly created duplicates rapidly go through neutral substitutions, during which they may be fixed or destroyed (pseudogenized or lost) due to genetic drift. After the fixation phase, the duplicates enter preservation phase, during which natural selection determines whether the duplicates can be retained.[63] Generally, the neofunctionalization model suggests that each of two originally identical gene copies can be retained if at least one evolves a new function.[8] The subfunctionalization model, which proposes that duplicated gene copies might both be retained if they subdivide the functions of the ancestral gene—for example, through complementary loss of different *cis* regulatory

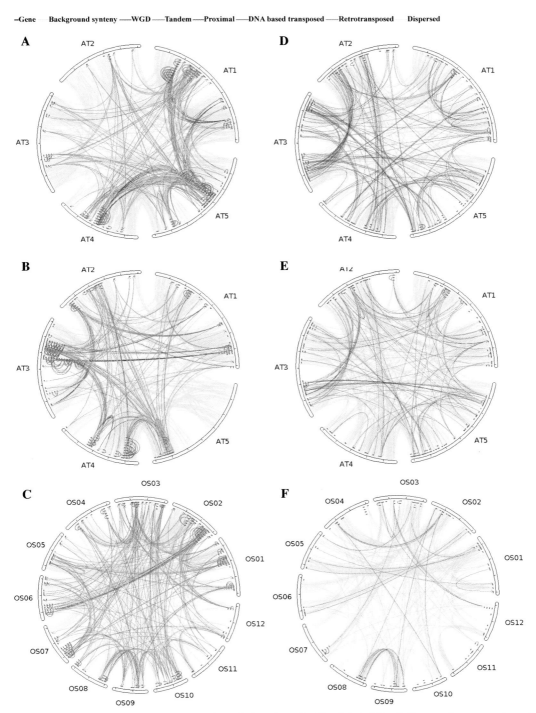

Figure 1. Gene duplication modes among the members of selected gene families. (A) *Arabidopsis* disease resistance gene homologs. (B) *Arabidopsis* cytochrome P450 gene family. (C) Rice cytochrome P450 gene family. (D) *Arabidopsis* cytoplasmic ribosomal gene family. (E) *Arabidopsis* C2H2 zinc finger gene family. (F) Rice C2H2 zinc finger gene family. Different gene duplication modes are indicated by different colors. From work by Wang *et al.*[62]

elements—has described an important modification of the neofunctionalization model.[9,69]

Some studies also show evidence to support the value of genetic redundancy per se.[10,12–17] Genetic redundancy is a common phenomenon in *Arabidopsis*, which may contribute to the absence of phenotypic effects in the majority of single loss-of-function mutants.[70] Genetic redundancy may lead to functional buffering or functional compensation. Immediately after gene duplication, duplicated copies are presumably fully redundant. Thus, loss of either copy will not result in a mutant phenotype. Full redundancy should be genetically unstable because the accumulation of deleterious mutations can result in rapid loss of one duplicate gene.[67,68] However, partial redundancy—that is, knockout of either duplicated copy resulting in more mitigated phenotypic changes than knockout of both copies—can be widespread in *Arabidopsis* and appears to be evolutionarily stable.[70] Phenotypic data for 5,041 *Arabidopsis* insertional mutant lines suggested that duplicate genes showed a significantly lower proportion of knockout effects than singleton genes.[71] Further, functional compensation by duplicate genes for severe phenotypic effects may tend to be preserved by natural selection for a longer time than for a less severe effect.[71] Chapman *et al.*[10] performed a genome-wide analysis of retention of WGD-duplicated genes in *Arabidopsis* and rice through analyzing SNP data. Contrary to classical predictions that duplicated genes should be relatively free to acquire unique functionality, they found that SNPs encode less radical amino acid changes in genes for which there exists a duplicated copy at a "paleologous" locus than in "singleton" genes. Preferential retention of duplicated genes that encode long complex proteins and their unexpectedly slow divergence suggest that a primary advantage of retaining WGD duplicates may be the buffering of crucial functions. Since functional buffering and functional divergence represent opposite extremes in the spectrum of duplicated gene fates, the authors suggested that functional buffering could be especially important during "genomic turmoil" immediately after genome duplication but might continue to act for tens of millions of years, and its gradual deterioration might contribute cyclicality to genome duplication in angiosperms.

We note that although functional buffering between duplicates can exist for tens of millions of years, this fact may not mean that natural selection is acting to favor its existence, especially from the point of view of population genetics. There are studies showing that selection for mutational robustness is necessarily very weak,[63,72–75] because the fitness advantage it confers can be no larger than the mutation rate. Innan *et al.*[63] suggested that under the model of shielding against deleterious mutations (i.e., functional buffering), extra duplicates are quickly destroyed by mutations, and therefore duplications must happen cyclically to provide sustained sheltering against deleterious mutations. Thus, duplicates may not be retained ultimately due to functional buffering.[63] In addition, the lower rates of nonsynonymous substitution in retained duplicates than in singletons were suspected to be due to the fact that a slow rate of evolution correlates with other factors, predisposing genes to be preserved in duplicates such as the number of regulatory regions or protein–protein interactions, or high levels of expression,[76–78] which can be understood as another model for duplicate gene retention: retention by gene conservation.[76,77]

Dosage balance theory has been related to retention of duplicated genes.[34,79–81] Veitia[82] stated the hypothesis that stoichiometric imbalance in macromolecular complexes can be a source of genetically dominant phenotypes. According to the hypothesis, Thomas *et al.*[81] suggested that a gene displays dosage effects increasingly as the subunit–subunit interactions of its product increase or from interactions with products downstream in a regulatory cascade, with positive and negative regulatory effectors. They defined "connected genes," as those most frequently interacting with other genes in a network. In 2009, Freeling[36] suggested that each mode of gene duplication often retains genes in a biased manner. WGD often retains genes related to transcription factors, protein kinases, and ribosomal proteins, while genes pertaining to DNA metabolism, defense proteins, and tRNA ligation are often lost following WGD. Potential enrichment of functional categories following tandem duplications was also reported. Tandem array genes may account for >10% of *Arabidopsis* or rice genes.[54] The comparison of properties of tandem genes between *Arabidopsis* and rice suggested that in both genomes, tandem arrays are enriched for genes that encode membrane proteins and function in "abiotic and biotic stress" but underrepresented for genes involved in transcription and DNA or RNA binding functions.[54] Further, some gene families are more likely to have transposed than other—for example, F-box, MADS genes, LRR-type disease resistance genes, and defensins.[43,83]

A general model for gene retention such as neofunctionalization or subfunctionalization may not fully explain the bias in plant gene content following different modes of gene duplication. The gene balance hypothesis appears more appropriate to account for the data of biased gene content.[36] The gene balance hypothesis postulates, according to Freeling,[36] that any successful genome has evolved an optimal balance of gene products that bind with one another to form protein complexes, or are involved in multiple steps of biological pathways/processes. In terms of network connectivity, it is hypothesized that the more connected the gene product, the more likely that the phenotype will change if dosage imbalance happens.[36] WGD duplicates all nuclear genes at once, so dosage relationships among genes do not change immediately. However, following WGD, many duplicates begin to revert to singleton status (i.e., fractionate). If connected duplicate genes fractionate, dosage imbalance is expected and may be harmful to the organism. Accordingly, duplicate genes that have been retained after WGD are more likely to be connected. In contrast, tandem duplication creates just one more duplicate gene. If a tandem duplicate is highly connected, it has a higher chance of causing dosage imbalance and thus is very likely to be removed. If the tandem duplicate tends to work alone, dosage imbalance is less likely to happen and thus the tandem duplicate has a higher probability of being retained.

The gene balance hypothesis appears consistent with the biased gene content following different modes of gene duplication. In addition, some studies provided data that in many gene families/functional categories, the frequencies of gene retention after WGD and tandem duplications are often reciprocal.[36,55,81] For example, Freeling[36] found that there are reciprocal relationships between *Arabidopsis* genes retained posttandem duplications versus posttetraploidy in the GO terms "transcription factor activity" and "structural constituent of the ribosome." He further showed the reciprocal relationships between WGD and tandem duplications among 16 transcription factor gene families with ≥30 genes, and concluded that only balanced gene drive predicts

the reciprocal relationships between WGD and tandem duplications among three evolutionary models involving neofunctionalization, subfunctionalization, and balanced gene drive. However, the gene balance hypothesis does not describe the long-term consequences of gene duplication—for example, functional divergence or innovations.

Another model for retention of duplicate genes relates to gene expression levels. This model was originally revealed through analyzing the genome of *Paramecium tetraurelia*, where a strong correlation was observed between expression levels and retention rates of WGD paralogs.[3] This model is compatible with the gene balance hypothesis, evidenced by a strong over-retention of genes involved in known complexes. Bekaert *et al.*[84] showed how these two models interplay in the *Arabidopsis* genome through analyzing properties of metabolic networks. They claimed that retention of WGD duplicates may result from both relative and absolute gene dosage constraints, representing the selections on the relative dosages of central network genes (i.e., the dosage balance hypothesis) and absolute increase in the concentration of gene products, respectively. They found that the retained duplicates from the α WGD event are clustered in the network, supporting the dosage balance hypothesis. However, the retained duplicates from the β WGD event are associated with high metabolic flux. They concluded that relative dosage constraints mainly affect the more recent WGD event and are being resolved by evolution and may be fully resolved for more ancient WGD events.

Researchers have often tried to find evidence in favor of a particular duplicate gene retention model over others. It appears that each model can be supported by some datasets, in particular lineages. However, along with increased knowledge of duplicate gene retention, it becomes clearer that duplicate gene retention can be a very complex biological process that can be affected by multiple factors. Indeed, it might be impossible or unnecessary to distinguish which model is the ultimate driving force. For example, the functional buffering model can be compatible with the gene balance hypothesis. Highly connected duplicate genes are more likely to be retained after WGD due to the need of dosage balance, but highly connected duplicate genes are also more likely to be involved in essential functions most likely to benefit from buffering. Further, it may be helpful to distinguish the evolutionary stage when a model functions. It is possible that the gene balance hypothesis and functional buffering model influence only the evolution of recently duplicated genes while their effects decay as time passes, but neofunctionalization is the long-term evolutionary trajectory. Furthermore, it remains unclear whether duplicate genes created by different mechanisms differ in retention and evolutionary models. In addition, it has been suggested that gene conversion can play essential and diverse roles in the models of duplicate gene retention.[63] The roles of epigenetic factors in duplicate gene retention are little understood and warrant future investigation.

Gene expression divergence following genome and gene duplications

Gene expression divergence after WGDs

The study of gene expression divergence can facilitate the understanding of evolutionary consequences of WGDs. In 2004, Blanc and Wolfe[85] published a study of gene expression divergence of WGD-derived duplicated *Arabidopsis* genes. They analyzed 1137 and 420 duplicated pairs formed by the most recent polyploidy event and the older polyploidy events, respectively. They found that 57% (653) of the pairs of more recent duplicates and 73% (306) of the pairs of older duplicates were diverged in expression. Further, they found 30 distinct pairs of recent duplicates showing concerted expression divergence. Furthermore, 21% (173) of 833 pairs of young duplicates showed asymmetric evolution in protein sequences. They concluded that 62% of the recent duplicate pairs presented evidence for functional diversification. In 2009, Throude *et al.*[86] conducted a structural and expression analysis of rice paleo-duplications. They identified 115 WGD duplicate pairs for which at least one copy was differentially expressed in one of three tissues, including grain, leaf, and root, finding that the vast majority of the 115 paralogous gene pairs had been neofunctionalized or subfunctionalized—that is, 88%, 89%, and 96% of duplicates expressed in grain, leaf, and root, respectively, showed distinct expression patterns. They concluded that the vast majority (>85%) of duplicates had been either lost or had been subfunctionalized or neofunctionalized during 50–70 million years of evolution. In 2009, Yim *et al.*[87] examined the evolu-

tionary dynamics of rice duplicated genes formed by paleopolyploidy prior to the radiation of the *Poaceae* family. They found that 57.4% of ~70 million years ago (MYA) duplicated genes and 50.9% of young ~7.7 MYA duplicated genes have diverged in expression. Recently, we found that in both *Arabidopsis* and rice, more ancient WGD-duplicated genes are likely to have greater expression divergence than more recent ones.[62] In partial summary, these studies reveal that in both *Arabidopsis* and rice, more than one half of duplicated genes formed by paleopolyploidy have diverged in gene expression.

Gene expression divergence in synthetic polyploids

Synthetic polyploidy has been a widely used platform to study the dynamic changes in genomic structures and gene expression caused by polyploidy. Merging and doubling of two genomes can stimulate extensive modifications of the genome and/or transcriptome, which create cascades of novel expression patterns, regulatory interactions, and new phenotypic variations that subsequent natural selection may act upon.[88–92] Most studies about gene expression changes following synthetic polyploidy have been based on allopolyploids (i.e., merging two different genomes),[93–100] although a few involve autopolyploids (doubling the same genome).[101–104] Several major findings with regard to allopolyploids can be summarized as follows. First, homologous genes may not contribute equally to the transcript pool. The studies on gene expression evolution in natural allopolyploid cotton, *A. suecica*, and wheat found numerous homologs that show silencing of one copy or strong expression bias toward one copy.[30,99,105] Second, many genes do not express as additive combinations of the parental genomes.[95,96] For example, in a synthetic *Arabidopsis* allotetraploid that was formed from *A. arenosa* and *A. thaliana*, genes showed strong expression dominance of the *A. arenosa* parent and most of the repressed genes were those that were upregulated in the *A. thaliana* parent relative to the *A. arenosa* parent.[96] Third, some genes may show repeatable silencing while others may show stochastic silencing.[100,106] Fourth, patterns of gene silencing may be organ specific.[100] Fifth, changes in gene expression are more affected by interspecific hybridization than by genome doubling.[97,107]

In contrast to the case in allopolyploids, where interspecific hybridization may trigger substantial changes of gene expression, genome doubling by autopolyploidy can restore a state similar to that of its diploid progenitors,[95] with genomic changes that are more moderate. In 2005, Albertin *et al.*[104] reported that genome doubling itself may not induce major changes of gene expression in *B. oleracea.* The consequences of synthetic autopolyploidy have been discussed in detail.[108]

Note that it remains unclear whether the dramatic changes associated with synthetic polyploidy are the beginning of adaptation to the newly duplicated state, or symptoms of imminent extinction of the lineage. Thus, neither synthetic autopolyploidy nor allopolyploidy may reveal the exact evolutionary dynamics or consequences of paleopolyploidy, as ancient environments cannot be recovered and selection and adaptation have worked for a long time.

Comparison of gene expression divergence among gene duplication modes

Gene duplication modes can affect expression divergence between duplicated genes. In 2006, Casneuf *et al.*[109] reported that in *Arabidopsis*, duplicated genes that were created by large-scale duplication events or that can be found in collinear segments have expression patterns that are more correlated with one another than those created by small-scale duplications or that are no longer located within collinear segments. They further found that the WGD duplicates tend to have highly redundant or overlapping expression patterns and are mostly expressed in the same tissues, while the other duplicates tend to show asymmetric divergence. In 2007, Ganko *et al.*[56] classified gene duplication modes in *Arabidopsis* into segmental, tandem, and dispersed duplications. They found that much of the expression divergence among tandem or dispersed duplicates may occur shortly after duplication, and polyploidy-derived duplicated genes have broader expression patterns and higher expression levels than nonpolyploidy-derived duplicated genes. In 2009, Li *et al.*[110] assessed the expression divergence of rice duplicated genes formed by block, tandem, and dispersed duplications. They found that expression correlation is significantly higher for duplicated genes from block and tandem duplications than those from dispersed duplications.

More recently, we summarized the expression levels of *Arabidopsis* and rice duplicate genes across 4,566 *Arabidopsis* and 508 rice microarrays, respectively.[62] Then we systematically compared expression divergence between genes duplicated by six different mechanisms including WGD, tandem, proximal, DNA-based transposed, retrotransposed, and dispersed (Fig. 2). It was found that the trends of expression divergence between duplicates in *Arabidopsis* and rice are very similar: DNA-based transposed duplication ≈ retrotransposed duplication > dispersed duplication > proximal duplication > WGD ≈ tandem duplication. This comparison suggests that, although most of duplicate genes have diverged in expression, the origins of genetic novelty, of clear biological significance in occupation of new niches or adaptation to new environments, may lie more with the greater expression divergence of distantly transposed and dispersed duplications.

The distinct evolutionary consequences following different modes of gene duplication suggests that the models dominating duplicate gene retention and evolution might not be unique. We have proposed a hypothesis.[62] WGD is often associated with speciation in plants.[111,112] If paleopolyploidy was attendant with speciation, new species would have likely initially faced very small *Ne* (i.e., effective population size), weak selection, high drift, and high mutational load, which could put a premium on buffering but allow little chance for beneficial mutations. On the other hand, transposed/dispersed duplications, which may have been only infrequently associated with speciation, might be more likely to arise in established populations with larger *Ne* and more efficient selection, all putting a greater premium on evolutionary novelty to attain fixation. If this hypothesis were true, WGD and small-scale duplicates might be retained under different evolutionary models due to different population sizes. Surely, the relationships between gene duplicate modes and models of duplicate gene retention have not been well elucidated and warrant future investigation.

Comparison of gene expression divergence across species

The study of gene expression divergence between *Drosophila* species and strains suggested that both within a genome and between genomes, duplicated genes are more likely to show changes in expression profiles than singletons.[113] In *Arabidopsis*, such a

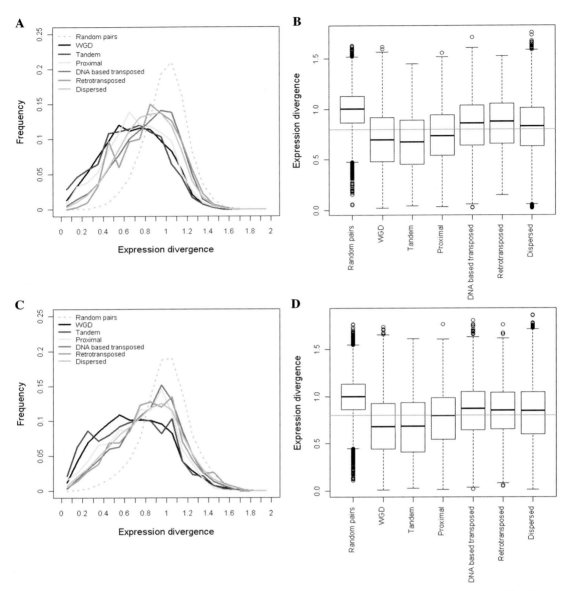

Figure 2. Comparison of expression divergence among different modes of gene duplication. (A) Comparison of distributions of expression divergence in *Arabidopsis*. (B) Comparison of levels of expression divergence in *Arabidopsis*. (C) Comparison of distributions of expression divergence in rice. (D) Comparison of levels of expression divergence in rice. Expression divergence between duplicate genes is measured by 1-*r*, where *r* is the correlation between their expression profiles. Green lines in B and D indicate average expression divergence across duplication modes. From work by Wang *et al.*[62]

tendency was examined and proven by Ha *et al.*[114] This finding suggests that gene duplication is a primary mechanism for increasing functional diversification, and the increased expression divergence in duplicated genes can substantially contribute to morphological diversification.[115,116]

The factors that affect expression divergence are complex

Gene expression divergence in eukaryotes has been often related to the dynamics of many genetic and epigenetic factors in addition to the underlying gene duplication mechanisms. This question has been also examined in angiosperms. First, gene expression divergence is related to the evolution of coding sequences such as synonymous (Ks) and nonsynonymous (Ka) nucleotide substitution rates. Casneuf *et al.*[109] analyzed how expression correlation between *Arabidopsis* duplicates changed with Ks in major functional categories and found that in general, the expression correlation was high for recently duplicated genes and declined with increasing time since duplication, but the correlation pattern differs among functional categories. Ganko *et al.*[56] suggested that there might be little correlation between expression divergence and Ks, whereas there could be a strong positive correlation between expression divergence and Ka, indicating that the strength of purifying selection acting on protein sequences and expression patterns was correlated. Li *et al.*[110] reported that in rice, there was a significant correlation between expression divergence and Ks for duplicated genes, and both duplication types and divergence time influenced the difference in expression divergence. Further, the difference in the extent of gene expression divergence between large and small-scale gene duplication was suspected to be related to different propensity for promoter disruption—that is, single-gene duplications that are often caused by unequal cross-over or duplicative transposition are much more prone to promoter disruption than genes duplicated through polyploidy events.[109] Recently, we reported that the extent of expression divergence between duplicates is discernibly related to duplication modes, different WGD events, amino acid divergence, and putatively neutral divergence (time), but the contribution of each factor is heterogeneous among duplication modes.[62]

In synthetic allopolyploids, chromosomal modification such as fragmentation and/or rearrangement may happen rapidly.[92,117] Additionally, transposons are often activated in newly synthetic polyploids.[32,118] These genetic changes could lead to expression divergence between duplicated genes. A recent study suggested that transposable elements and small RNAs contributed to gene expression divergence both within and between *Arabidopsis thaliana* and *Arabidopsis lyrata*.[119] In addition, many studies have suggested that epigenetic changes such as DNA methylation and histone modification are often associated with gene expression divergence in polyploids.[32,88,89,92,117] Findings to date suggest that both genetic and epigenetic factors can be related to gene expression divergence and that gene expression divergence itself might be caused by the interaction of multiple genetic and epigenetic factors. New factors that may affect expression divergence and how different factors work together warrant further investigation.

Conclusions

Angiosperms are an outstanding model for studying and elucidating the evolutionary mechanisms and consequences of genome and gene duplications. Genes can be duplicated by at least six different mechanisms, including whole-genome, tandem, proximal, DNA-based transposed, retrotransposed, and dispersed duplications. Besides now-classical neofunctionalizaiton and subfunctionalization models, various evolutionary models, such as functional buffering and the gene balance hypothesis, have been proposed to explain biased duplicate gene retention. Duplicate genes tend to diverge in expression in paleopolyploids, and the levels of expression divergence differ among gene duplication modes. Promising future directions are to exactly depict different mechanisms of gene duplication and develop models to explain distinct evolutionary consequences following different modes of gene duplication.

Acknowledgments

A.H.P. appreciates funding from the National Science Foundation (NSF: DBI 0849896, MCB 0821096, MCB 1021718). We thank Xu Tan and Hui Guo for helpful discussion and Barry Marler for IT support.

Conflicts of interest

The authors declare no conflicts of interest.

References

1. Paterson, A.H. *et al.* 2010. Insights from the comparison of plant genome sequences. *Annu. Rev. Plant. Biol.* **61:** 349–372.
2. Jaillon, O. *et al.* 2004. Genome duplication in the teleost fish *Tetraodon nigroviridis* reveals the early vertebrate protokaryotype. *Nature* **431:** 946–957.
3. Aury, J.M. *et al.* 2006. Global trends of whole-genome duplications revealed by the ciliate *Paramecium tetraurelia. Nature* **444:** 171–178.
4. Wolfe, K.H. & D.C. Shields. 1997. Molecular evidence for an ancient duplication of the entire yeast genome. *Nature* **387:** 708–713.

5. Kellis, M., B.W. Birren & E.S. Lander. 2004. Proof and evolutionary analysis of ancient genome duplication in the yeast *Saccharomyces cerevisiae*. *Nature* **428:** 617–624.

6. Kassahn, K.S. *et al.* 2009. Evolution of gene function and regulatory control after whole-genome duplication: comparative analyses in vertebrates. *Genome. Res.* **19:** 1404–1418.

7. Zhang, G. & M.J. Cohn. 2008. Genome duplication and the origin of the vertebrate skeleton. *Curr. Opin. Genet. Dev.* **18:** 387–393.

8. Ohno, S. 1970. *Evolution by gene duplication.* Springer Verlag. New York.

9. Lynch, M. & J.S. Conery. 2000. The evolutionary fate and consequences of duplicate genes. *Science* **290:** 1151–1155.

10. Chapman, B.A. *et al.* 2006. Buffering of crucial functions by paleologous duplicated genes may contribute cyclicality to angiosperm genome duplication. *Proc. Natl. Acad. Sci. USA* **103:** 2730–2735.

11. VanderSluis, B. *et al.* 2010. Genetic interactions reveal the evolutionary trajectories of duplicate genes. *Mol. Syst. Biol.* **6:** 429.

12. Dean, E.J. *et al.* 2008. Pervasive and persistent redundancy among duplicated genes in yeast. *PLoS Genet.* **4:** e1000113.

13. DeLuna, A. *et al.* 2010. Need-based up-regulation of protein levels in response to deletion of their duplicate genes. *PLoS Biol.* **8:** e1000347.

14. DeLuna, A. *et al.* 2008. Exposing the fitness contribution of duplicated genes. *Nat. Genet.* **40:** 676–681.

15. Gu, Z. *et al.* 2003. Role of duplicate genes in genetic robustness against null mutations. *Nature* **421:** 63–66.

16. Kafri, R. *et al.* 2008. Preferential protection of protein interaction network hubs in yeast: evolved functionality of genetic redundancy. *Proc. Natl. Acad. Sci. USA* **105:** 1243–1248.

17. Emili, A. *et al.* 2008. The extensive and condition-dependent nature of epistasis among whole-genome duplicates in yeast. *Genome. Res.* **18:** 1092–1099.

18. Wang, X. *et al.* 2009. Comparative inference of illegitimate recombination between rice and sorghum duplicated genes produced by polyploidization. *Genome. Res.* **19:** 1026–1032.

19. Wang, X., H. Tang & A.H. Paterson. 2011. Seventy million years of concerted evolution of a homoeologous chromosome pair, in parallel, in major Poaceae lineages. *Plant Cell.* **23:** 27–37.

20. Wang, X. *et al.* 2007. Extensive concerted evolution of rice paralogs and the road to regaining independence. *Genetics* **177:** 1753–1763.

21. Song, K. *et al.* 1995. Rapid genome change in synthetic polyploids of *Brassica* and its implications for polyploid evolution. *Proc. Natl. Acad. Sci. USA* **92:** 7719–7723.

22. Ozkan, H., A.A. Levy & M. Feldman. 2001. Allopolyploidy-induced rapid genome evolution in the wheat (*Aegilops-Triticum*) group. *Plant Cell.* **13:** 1735–1747.

23. Shaked, H. *et al.* 2001. Sequence elimination and cytosine methylation are rapid and reproducible responses of the genome to wide hybridization and allopolyploidy in wheat. *Plant Cell.* **13:** 1749–1759.

24. Kashkush, K., M. Feldman & A.A. Levy. 2002. Gene loss, silencing and activation in a newly synthesized wheat allotetraploid. *Genetics* **160:** 1651–1659.

25. Kashkush, K., M. Feldman & A.A. Levy. 2003. Transcriptional activation of retrotransposons alters the expression of adjacent genes in wheat. *Nat. Genet.* **33:** 102–106.

26. O'Neill, R.J., M.J. O'Neill & J.A. Graves. 1998. Undermethylation associated with retroelement activation and chromosome remodelling in an interspecific mammalian hybrid. *Nature* **393:** 68–72.

27. Paterson, A.H. *et al.* 2009. The *Sorghum bicolor* genome and the diversification of grasses. *Nature* **457:** 551–556.

28. Chen, Z.J. & C.S. Pikaard. 1997. Transcriptional analysis of nucleolar dominance in polyploid plants: biased expression/silencing of progenitor rRNA genes is developmentally regulated in *Brassica*. *Proc. Natl. Acad. Sci. USA* **94:** 3442–3447.

29. Comai, L. *et al.* 2000. Phenotypic instability and rapid gene silencing in newly formed *Arabidopsis* allotetraploids. *Plant Cell.* **12:** 1551–1568.

30. Lee, H.S. & Z.J. Chen. 2001. Protein-coding genes are epigenetically regulated in *Arabidopsis* polyploids. *Proc. Natl. Acad. Sci. USA* **98:** 6753–6758.

31. Rodin, S.N. & A.D. Riggs. 2003. Epigenetic silencing may aid evolution by gene duplication. *J. Mol. Evol.* **56:** 718–729.

32. Adams, K.L. & J.F. Wendel. 2005. Novel patterns of gene expression in polyploid plants. *Trends Genet.* **21:** 539–543.

33. Rapp, R.A. & J.F. Wendel. 2005. Epigenetics and plant evolution. *New Phytol.* **168:** 81–91.

34. Maere, S. *et al.* 2005. Modeling gene and genome duplications in eukaryotes. *Proc. Natl. Acad. Sci. USA* **102:** 5454–5459.

35. Cusack, B.P. & K.H. Wolfe. 2007. Not born equal: increased rate asymmetry in relocated and retrotransposed rodent gene duplicates. *Mol. Biol. Evol.* **24:** 679–686.

36. Freeling, M. 2009. Bias in plant gene content following different sorts of duplication: tandem, whole-genome, segmental, or by transposition. *Annu. Rev. Plant Biol.* **60:** 433–453.

37. Bowers, J.E. *et al.* 2003. Unravelling angiosperm genome evolution by phylogenetic analysis of chromosomal duplication events. *Nature* **422:** 433–438.

38. Paterson, A.H., J.E. Bowers & B.A. Chapman. 2004. Ancient polyploidization predating divergence of the cereals, and its consequences for comparative genomics. *Proc. Natl. Acad. Sci. USA* **101:** 9903–9908.

39. Tang, H. *et al.* 2008. Synteny and collinearity in plant genomes. *Science* **320:** 486–488.

40. Tang, H. *et al.* 2008. Unraveling ancient hexaploidy through multiply-aligned angiosperm gene maps. *Genome. Res.* **18:** 1944–1954.

41. Tang, H. *et al.* 2010. Angiosperm genome comparisons reveal early polyploidy in the monocot lineage. *Proc. Natl. Acad. Sci. USA* **107:** 472–477.

42. Jiao, Y. *et al.* 2011. Ancestral polyploidy in seed plants and angiosperms. *Nature* **473:** 97–100.

43. Freeling, M. *et al.* 2008. Many or most genes in *Arabidopsis* transposed after the origin of the order Brassicales. *Genome. Res.* **18:** 1924–1937.

44. Coghlan, A. *et al.* 2005. Chromosome evolution in eukaryotes: a multi-kingdom perspective. *Trends Genet.* **21:** 673–682.

45. Van de Peer, Y. 2004. Computational approaches to unveiling ancient genome duplications. *Nat. Rev. Genet.* **5:** 752–763.

46. Vandepoele, K. *et al.* 2002. The automatic detection of homologous regions (ADHoRe) and its application to microcolinearity between *Arabidopsis* and rice. *Genome. Res.* **12:** 1792–1801.

47. Cannon, S.B. *et al.* 2003. DiagHunter and GenoPix2D: programs for genomic comparisons, large-scale homology discovery and visualization. *Genome. Biol.* **4:** R68.

48. Calabrese, P.P., S. Chakravarty & T.J. Vision. 2003. Fast identification and statistical evaluation of segmental homologies in comparative maps. *Bioinformatics* **19:** i74–i80.

49. Haas, B.J. *et al.* 2004. DAGchainer: a tool for mining segmental genome duplications and synteny. *Bioinformatics* **20:** 3643–3646.

50. Wang, X. *et al.* 2006. Statistical inference of chromosomal homology based on gene colinearity and applications to *Arabidopsis* and rice. *BMC Bioinformatics* **7:** 447.

51. Vandepoele, K., C. Simillion & Y. Van de Peer. 2002. Detecting the undetectable: uncovering duplicated segments in *Arabidopsis* by comparison with rice. *Trends Genet.* **18:** 606–608.

52. Vandepoele, K., C. Simillion & Y. Van de Peer. 2003. Evidence that rice and other cereals are ancient aneuploids. *Plant Cell.* **15:** 2192–2202.

53. Dujon, B. *et al.* 2004. Genome evolution in yeasts. *Nature* **430:** 35–44.

54. Gaut, B.S., C. Rizzon & L. Ponger. 2006. Striking similarities in the genomic distribution of tandemly arrayed genes in *Arabidopsis* and rice. *PLoS Comput. Biol.* **2:** e115.

55. Cannon, S.B. *et al.* 2004. The roles of segmental and tandem gene duplication in the evolution of large gene families in *Arabidopsis thaliana*. *BMC Plant Biol.* **4:** 10.

56. Ganko, E.W., B.C. Meyers & T.J. Vision. 2007. Divergence in expression between duplicated genes in *Arabidopsis*. *Mol. Biol. Evol.* **24:** 2298–2309.

57. Jiang, N. *et al.* 2004. Pack-MULE transposable elements mediate gene evolution in plants. *Nature* **431:** 569–573.

58. Brunner, S. *et al.* 2005. Evolution of DNA sequence nonhomologies among maize inbreds. *Plant Cell.* **17:** 343–360.

59. Kaessmann, H., N. Vinckenbosch & M. Long. 2009. RNA-based gene duplication: mechanistic and evolutionary insights. *Nat. Rev. Genet.* **10:** 19–31.

60. Brosius, J. 1991. Retroposons—seeds of evolution. *Science* **251:** 753.

61. Zhao, X.P. *et al.* 1998. Dispersed repetitive DNA has spread to new genomes since polyploid formation in cotton. *Genome Res.* **8:** 479–492.

62. Wang, Y. *et al.* 2011. Modes of gene duplication contribute differently to genetic novelty and redundancy, but show parallels across divergent angiosperms. *PLoS One* **6:** e28150.

63. Innan, H. & F. Kondrashov. 2010. The evolution of gene duplications: classifying and distinguishing between models. *Nat. Rev. Genet.* **11:** 97–108.

64. Conant, G.C. & K.H. Wolfe. 2008. Turning a hobby into a job: how duplicated genes find new functions. *Nat. Rev. Genet.* **9:** 938–950.

65. Innan, H. 2009. Population genetic models of duplicated genes. *Genetica.* **137:** 19–37.

66. Hahn, M.W. 2009. Distinguishing among evolutionary models for the maintenance of gene duplicates. *J. Hered.* **100:** 605–617.

67. Wagner, A. 1998. The fate of duplicated genes: loss or new function? *Bioessays.* **20:** 785–788.

68. Tautz, D. 1992. Redundancies, development and the flow of information. *Bioessays.* **14:** 263–266.

69. Force, A. *et al.* 1999. Preservation of duplicate genes by complementary, degenerative mutations. *Genetics* **151:** 1531–1545.

70. Briggs, G.C. *et al.* 2006. Unequal genetic redundancies in *Arabidopsis*—a neglected phenomenon? *Trends Plant Sci.* **11:** 492–498.

71. Hanada, K. *et al.* 2009. Evolutionary persistence of functional compensation by duplicate genes in *Arabidopsis*. *Genome. Biol. Evol.* **1:** 409–414.

72. Cooke, J. *et al.* 1997. Evolutionary origins and maintenance of redundant gene expression during metazoan development. *Trends Genet.* **13:** 360–364.

73. Nowak, M.A. *et al.* 1997. Evolution of genetic redundancy. *Nature* **388:** 167–171.

74. Wagner, A. 1999. Redundant gene functions and natural selection. *J. Evol. Biol.* **12:** 1–16.

75. Wagner, A. 2000. The role of population size, pleiotropy and fitness effects of mutations in the evolution of overlapping gene functions. *Genetics* **154:** 1389–1401.

76. Davis, J.C. & D.A. Petrov. 2004. Preferential duplication of conserved proteins in eukaryotic genomes. *PLoS Biol.* **2:** E55.

77. Davis, J.C. & D.A. Petrov. 2005. Do disparate mechanisms of duplication add similar genes to the genome? *Trends Genet.* **21:** 548–551.

78. Semon, M. & K.H. Wolfe. 2007. Consequences of genome duplication. *Curr. Opin. Genet. Dev.* **17:** 505–512.

79. Veitia, R.A. 2003. Nonlinear effects in macromolecular assembly and dosage sensitivity. *J. Theor. Biol.* **220:** 19–25.

80. Papp, B., C. Pal & L.D. Hurst. 2003. Dosage sensitivity and the evolution of gene families in yeast. *Nature* **424:** 194–197.

81. Thomas, B.C. & M. Freeling. 2006. Gene-balanced duplications, like tetraploidy, provide predictable drive to increase morphological complexity. *Genome. Res.* **16:** 805–814.

82. Veitia, R.A. 2004. Gene dosage balance in cellular pathways: implications for dominance and gene duplicability. *Genetics* **168:** 569–574.

83. Woodhouse, M.R., B. Pedersen & M. Freeling. 2010. Transposed genes in *Arabidopsis* are often associated with flanking repeats. *PLoS Genet.* **6:** e1000949.

84. Bekaert, M. *et al.* 2011. Two-phase resolution of polyploidy in the *Arabidopsis* metabolic network gives rise to relative and absolute dosage constraints. *Plant Cell.* **23:** 1719–1728.

85. Blanc, G. & K.H. Wolfe. 2004. Functional divergence of duplicated genes formed by polyploidy during *Arabidopsis* evolution. *Plant Cell.* **16:** 1679–1691.

86. Throude, M. *et al.* 2009. Structure and expression analysis of rice paleo duplications. *Nucleic Acids Res.* **37:** 1248–1259.

87. Yim, W.C., B.M. Lee & C.S. Jang. 2009. Expression diversity and evolutionary dynamics of rice duplicate genes. *Mol. Genet. Genomics* **281:** 483–493.

88. Adams, K.L. & J.F. Wendel. 2005. Polyploidy and genome evolution in plants. *Curr. Opin. Plant. Biol.* **8:** 135–141.

89. Osborn, T.C. *et al.* 2003. Understanding mechanisms of novel gene expression in polyploids. *Trends Genet.* **19:** 141–147.

90. Adams, K.L. 2007. Evolution of duplicate gene expression in polyploid and hybrid plants. *J. Hered.* **98:** 136–141.

91. Doyle, J.J. *et al.* 2008. Evolutionary genetics of genome merger and doubling in plants. *Annu. Rev. Genet.* **42:** 443–461.

92. Chen, Z.J. 2007. Genetic and epigenetic mechanisms for gene expression and phenotypic variation in plant polyploids. *Annu. Rev. Plant. Biol.* **58:** 377–406.

93. Rapp, R.A., J.A. Udall & J.F. Wendel. 2009. Genomic expression dominance in allopolyploids. *BMC Biol.* **7:** 18.

94. Flagel, L.E. & J.F. Wendel. 2010. Evolutionary rate variation, genomic dominance and duplicate gene expression evolution during allotetraploid cotton speciation. *New Phytol.* **186:** 184–193.

95. Hegarty, M.J. *et al.* 2006. Transcriptome shock after interspecific hybridization in senecio is ameliorated by genome duplication. *Curr. Biol.* **16:** 1652–1659.

96. Wang, J. *et al.* 2006. Genomewide nonadditive gene regulation in *Arabidopsis* allotetraploids. *Genetics* **172:** 507–517.

97. Flagel, L. *et al.* 2008. Duplicate gene expression in allopolyploid *Gossypium* reveals two temporally distinct phases of expression evolution. *BMC Biol.* **6:** 16.

98. Auger, D.L. *et al.* 2005. Nonadditive gene expression in diploid and triploid hybrids of maize. *Genetics* **169:** 389–397.

99. Adams, K.L. *et al.* 2003. Genes duplicated by polyploidy show unequal contributions to the transcriptome and organ-specific reciprocal silencing. *Proc. Natl. Acad. Sci. USA* **100:** 4649–4654.

100. Adams, K.L., R. Percifield & J.F. Wendel. 2004. Organ-specific silencing of duplicated genes in a newly synthesized cotton allotetraploid. *Genetics* **168:** 2217–2226.

101. Stupar, R.M. *et al.* 2007. Phenotypic and transcriptomic changes associated with potato autopolyploidization. *Genetics* **176:** 2055–2067.

102. Chen, W.S. *et al.* 2006. A genome-wide comparison of genes responsive to autopolyploidy in Isatis indigotica using *Arabidopsis thaliana* Affymetrix genechips. *Plant Mol. Biol. Rep.* **24:** 197–204.

103. Martelotto, L.G. *et al.* 2005. A comprehensive analysis of gene expression alterations in a newly synthesized *Paspalum notatum* autotetraploid. *Plant Sci.* **169:** 211–220.

104. Albertin, W. *et al.* 2005. Autopolyploidy in cabbage (*Brassica oleracea* L.) does not alter significantly the proteomes of green tissues. *Proteomics* **5:** 2131–2139.

105. Bottley, A., G.M. Xia & R.M. Koebner. 2006. Homoeologous gene silencing in hexaploid wheat. *Plant J.* **47:** 897–906.

106. Wang, J. *et al.* 2004. Stochastic and epigenetic changes of gene expression in *Arabidopsis* polyploids. *Genetics* **167:** 1961–1973.

107. Albertin, W. *et al.* 2006. Numerous and rapid nonstochastic modifications of gene products in newly synthesized *Brassica napus* allotetraploids. *Genetics* **173:** 1101–1113.

108. Parisod, C., R. Holderegger & C. Brochmann. 2010. Evolutionary consequences of autopolyploidy. *New Phytol.* **186:** 5–17.

109. Casneuf, T. *et al.* 2006. Nonrandom divergence of gene expression following gene and genome duplications in the flowering plant *Arabidopsis thaliana*. *Genome. Biol.* **7:** R13.

110. Li, Z. *et al.* 2009. Expression pattern divergence of duplicated genes in rice. *BMC Bioinformatics* **10**(Suppl 6): S8.

111. Stebbins, G.L. 1982. Plant speciation. *Prog. Clin. Biol. Res.* **96:** 21–39.

112. Wood, T.E. *et al.* 2009. The frequency of polyploid speciation in vascular plants. *Proc. Natl. Acad. Sci. USA* **106:** 13875–13879.

113. Gu, Z. *et al.* 2004. Duplicate genes increase gene expression diversity within and between species. *Nat Genet.* **36:** 577–579.

114. Ha, M., E.D. Kim & Z.J. Chen. 2009. Duplicate genes increase expression diversity in closely related species and allopolyploids. *Proc. Natl. Acad. Sci. USA* **106:** 2295–2300.

115. Hanada, K. *et al.* 2009. Increased expression and protein divergence in duplicate genes is associated with morphological diversification. *PLoS Genet.* **5:** e1000781.

116. Kliebenstein, D.J. 2008. A role for gene duplication and natural variation of gene expression in the evolution of metabolism. *PLoS One* **3:** e1838.

117. Jackson, S. & Z.J. Chen. 2010. Genomic and expression plasticity of polyploidy. *Curr. Opin. Plant. Biol.* **13:** 153–159.

118. Madlung, A. *et al.* 2005. Genomic changes in synthetic *Arabidopsis* polyploids. *Plant J.* **41:** 221–230.

119. Hollister, J.D. *et al.* 2011. Transposable elements and small RNAs contribute to gene expression divergence between *Arabidopsis thaliana* and *Arabidopsis lyrata*. *Proc. Natl. Acad. Sci. USA* **108:** 2322–2327.

Ann. N.Y. Acad. Sci. ISSN 0077-8923

Hox gene evolution: multiple mechanisms contributing to evolutionary novelties

Leslie Pick and Alison Heffer

Department of Entomology and Program in Molecular & Cell Biology, University of Maryland, College Park, Maryland

Address for correspondence: Leslie Pick, Department of Entomology and Program in Molecular & Cell Biology, 4112 Plant Sciences Building, University of Maryland, College Park, MD 20742. lpick @umd.edu

Hox genes, which are important for determining regional identity in organisms as diverse as flies and humans, are typically considered to be under strong evolutionary constraints because large changes in body plan are usually detrimental to survival. Despite this, there is great body plan diversity in nature, and many of the mechanisms underlying this diversity have been attributed to changes in *Hox* genes. Over the past year, several studies have examined how *Hox* genes play a role in evolution of body plans and novelties. Here, we examine four distinct evolutionary mechanisms implicated in *Hox* gene evolution, which include changes in (1) *Hox* gene expression, (2) downstream *Hox* target gene regulation without change in *Hox* expression, (3) protein-coding sequence, and (4) posttranscriptional regulation of *Hox* gene function. We discuss how these types of changes in *Hox* genes—once thought to be evolutionarily static—underlie morphological diversification. We review recent studies that highlight each of these mechanisms and discuss their roles in the evolution of morphology and novelties.

Keywords: *Hox* evolution; evo-devo; *ftz*; segmentation

Introduction

If one were to examine and compare a fly and then a mammal (e.g, human or mouse), one might see relatively little similarity in the overall appearance of these animals. Aside from the enormous differences in size, other key features differ substantially: flies have three pairs of legs, wings, hard exoskeletons, and bulging red eyes, whereas humans walk upright, mice on four legs, neither can fly, and they are covered by soft skin and fur, each to different extents. Yet certain aspects of the way in which the bodies of these animals are organized— known as the body plan—are similar among all three species: for example, they each have anterior and posterior ends, and each has bilateral symmetry. These latter features characterize all bilaterians, contributing to the evidence that animals such as flies and humans arose from a common ancestor, the Urbilaterian, more than 550 million years ago.[1] How did animals so different in form evolve from the Urbilaterian ancestor? Although we are still far from an answer to this question, the synthesis of the fields of developmental biology and genetics with that of evolutionary biology— Evo-Devo—has begun to provide insights into mechanisms underlying the diversification of body plans.

Since the 1930s, developmental geneticists such as Glücksohn–Waelsch and Waddington have sought to identify the genes controlling the development of the single fertilized egg cell into a complex, functioning organism. The novelty of their approach was the logic they used to probe this question. By examining animals carrying a mutation in a specific gene and comparing the phenotype to that of normal (wild type) animals, they inferred the function of the gene carrying the mutation (reviewed in Ref. 2). Much of their work was done in model organisms that could be easily reared in the laboratory, notably the mouse *Mus musculus* and the fruit fly *Drosophila melanogaster*. Although these pursuits shared a goal of identifying the genes that direct development of the body plan, it was presumed that the embryonic regulatory genes that control development would differ in different species, in keeping with the

doi: 10.1111/j.1749-6632.2011.06385.x

differences in body plan discussed above. However, the identification of sets of embryonic regulatory genes in *Drosophila* in the 1980s overturned this thinking and showed instead that a common genetic toolkit regulates the body plans of diverse metazoans (reviewed in Ref. 3).

Homeosis and the discovery of Hox *genes*

The term *homeosis* was coined and defined by Bateson in the late 1800s, who observed animals in nature that had one body part "changed into the likeness of" another, such as antennae transformed to legs (p. 85).[4] In his book describing these rather bizarre homeotic transformations, Bateson suggested that they are the basis of morphological evolution, an insight far ahead of its time. For years after this, these mutations remained a puzzle to scientists, and it was in *Drosophila* that their mechanism was revealed.

One of the most famous examples of a homeotic transformation is the four-winged fruit fly studied by Ed Lewis: here, the third thoracic (T3) segment, which normally lacks wings, is replaced by a second thoracic-like (T2) segment with a perfect pair of wings.[5–7] In another startling example—the Antennapedia (*Antp*) mutation—the antennae of the fly are replaced with a perfect pair of legs—the exact legs that would normally develop on the T2 segment.[8–16] Through years of study of these homeotic genes, it became clear that the normal or wild type function of these genes is to determine the unique identities of individual segments. For example, *Antp* normally specifies the unique identity of the T2 segment, including its specific leg. When *Antp* is misexpressed in the developing head, it does its job of patterning the T2 leg, but it does it in the wrong place, giving an adult fly with legs where the antennae should be.[17] Similarly, other homeotic genes specify other unique identities—for example, *Sex combs reduced* (*Scr*) specifies the identity of the leg on the first thoracic segment (T1), which, in males, bears specialized structures known as sex combs. Loss-of-function mutations in *Scr* thus lead to loss of T1-identity, evidenced by loss of sex combs.[12,13,18–20] Another feature that emerged from genetic analysis of homeotic genes was the idea proposed by Ed Lewis of colinearity.[21] *Drosophila* homeotic genes are located in a chromosomal cluster, and they affect body patterning from the anterior to the posterior of the animal in the order they are arranged within the complex (*HOM-C*, Fig. 1A and D).

How homeotic genes function to specify unique developmental pathways along the anterior–posterior body axis became clear, at least in principle, after these genes were sequenced.[22–25] Homeotic genes were found to share a short 180 base-pair region, termed the *homeobox*. These homeobox-containing genes have since been referred to as *Hox* genes. The homeobox encodes a DNA-binding domain, the *homeodomain*,[26,27] and *Hox* proteins function as transcription factors that regulate gene expression by binding to specific DNA sequences in *cis*-regulatory regions of a number of genes.[28–34] As such, they serve as master regulators or selector genes to initiate developmental programs by activating the expression of downstream or realizator genes involved in growth and differentiation of particular body structures.[30,35] For example, and broadly speaking, *Scr* would bind to *cis*-regulatory regions controlling genes involved in T1 identity and regulate their transcription, and *Antp* would regulate genes involved in T2 identity.

As these regulatory mechanisms were being unveiled through studies of *Drosophila*, it was found that the *Hox* genes are highly conserved throughout the animal kingdom in both structure and function.[24,33,36] *Hox* genes were shown to regulate the development of all kinds of organisms, from flies to mammals, as *Hox* mutations were also shown to cause homeotic transformations of the axial skeleton in mice (for an early example, see Ref. 37). Further, the colinear expression of *Hox* genes described above for *Drosophila* was conserved in mammals despite approximately 550 million years of evolutionary divergence (Fig. 1). Functional studies in which mammalian orthologs of *Drosophila Hox* genes were misexpressed in the fly showed that they retained the ability to generate fly-like homeotic transformations.[38–40] For example, when we expressed the mammalian ortholog of *Scr* (*Hox1.3*, renamed *Hoxa5*) in the fly, it transformed antennae into T1 legs, marked by sex combs—a bona fide fly homeotic transformation, mimicking the *Scr* homeotic phenotype,[41,42] although *Hoxa5* obviously plays a very different role in the mouse.[43–48]

These kinds of results changed developmental biologists' way of thinking: here were proteins that regulate the embryonic development of flies and mammals—animals with very different body

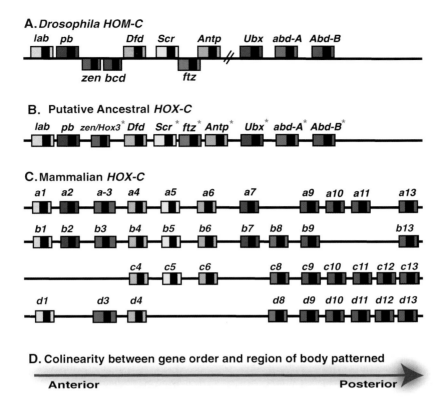

Figure 1. The anterior–posterior organization of *Hox* clusters is conserved from insects to mammals. (A) The *Drosophila Hox* complex (HOM-C) includes 11 genes, three of which have lost *Hox* function in higher insects (below the line: *zen, bcd, ftz*). (B) The putative ancestral *Hox* cluster included 10 genes. Evolutionary changes have been documented this year in the seven genes indicated by a red asterisk. (C) Mammals have four *Hox* clusters, *HoxA–D*. Gains and losses of paralogs have occurred within each cluster. (D) Colinearity of the order of the genes in *Hox* complexes and their function along the anterior–posterior body axis, conserved in flies and mammals, is indicated. *Hox* orthologs are colored the same (for example, *Drosophila Scr*, its mammalian orthologs, and their putative ancestor are shown in yellow). The black box represents the highly conserved homeobox.

types—yet they were conserved in sequence, structure, and even function. This raised a new set of questions in the field: rather than wondering if animals are similar enough that we can learn lessons about human biology from studying a fly, we now needed to ask how it could be that the embryonic development of diverse organisms is controlled by the same sets of genes, the genetic toolkit of animal development.[3] What is it that makes animal body plans different from each other?

Here, we discuss current thinking about the resolution of this *Hox* conservation paradox. We summarize the fundamental mechanisms that allow for evolutionary change in these key components of the genetic toolkit. We observe that changes in *Hox* genes are sometimes quite subtle, but even small changes may have significant effects on mor-

phology. Iterative rounds of small changes at the genetic level can produce new configurations of structures, gene networks, and body types that persist in nature.

How have Hox *genes changed during evolution?*

It is now widely accepted that Urbilateria harbored a *Hox* complex (Putative Ancestral *HoxC*, Fig. 1B) complete with most of the genes present in modern day *Hox* complexes.[49–51] In some lineages, duplication of entire *Hox* gene clusters has led to multiple copies of these genes, enabling diversification of function of individual paralogs, loss of paralogs because of redundancy, and additional gene duplications in some lineages.[52] This is evidenced in mammalian *Hox* clusters that have undergone

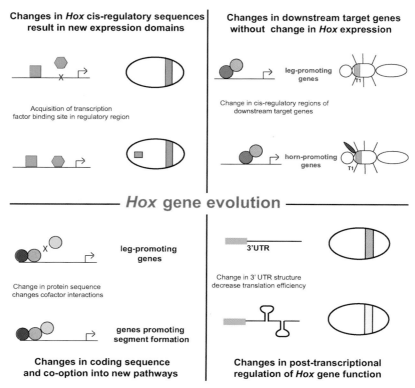

Figure 2. Four mechanisms underlie *Hox* regulatory evolution. A schematic of the four evolutionary mechanisms discussed in this review is shown. (Upper left) Changes in *cis*-regulatory regions. For example, a transcription factor (blue square) binds to a *cis*-regulatory element and directs *Hox* expression in a specific domain. A binding site for a new transcription factor (blue hexagon) is acquired, resulting in a novel expression domain. (Upper right) Changes in downstream target genes without a *Hox* expression change. For example, a *Hox* protein (blue circle) and its cofactor (orange circle) promote the activation of leg-specification genes. A change in *cis*-regulatory sequences of target genes occurs, such that the *Hox* protein and its partner now bind to and activate horn-specification genes. The expression of the *Hox* gene itself is unchanged. (Lower left) Changes in coding sequence and cooption into new pathways. For example, two transcription factors (purple grey circles) bind to DNA and activate genes that promote leg formation. If a mutation occurs in the *Hox* protein, such that it can now interact with a new cofactor (green circle), a new function may be acquired.(Lower right) Changes in posttranscriptional regulation of a *Hox* gene. For example, one 3′ UTR structure (indicated here as a straight line) is efficient in promoting translation of this protein, whereas another 3′ UTR structure (hairpins) results in less-efficient translation and a decrease in protein products. *Note:* mechanisms shown in the upper right and lower left panels each result in change in Hox function without affecting the expression of the *Hox* gene. Thus, when a *Hox* gene pattern is unchanged but new downstream functions are acquired, either mechanism may be at play.

two rounds of replication to generate four clusters (Fig. 1C).[33,53,54] Within each cluster, most genes are conserved, but some have been lost (e.g., *Hoxb10* and *Hoxc3*), whereas others expanded (e.g., the posterior *Hox* genes, represented only by *Abd-B* in *Drosophila*, have expanded in vertebrate lineages).

Other evolutionary changes documented to date in *Hox* function act upon the ancestral and conserved core of *Hox* genes thought to have been present in the Urbilateria. Some of these changes influenced the regulation and expression of *Hox* genes (*cis*-regulatory changes),[3,55,56] whereas others affected *Hox* protein activity (protein coding changes).[57–59] Still other changes occurred downstream of the *Hox* genes themselves, particularly in the regulatory regions of targets, which can be gained or lost in a lineage-specific fashion, thereby changing the biological role of a *Hox* gene without changes its expression. In all cases, gene regulatory networks (GRNs)[60–62] regulated by *Hox* genes are altered, although the mechanisms underlying this alteration are different. We have classified these mechanisms into four categories, diagramed in Figure 2: (1) changes in *Hox* gene expression,

(2) changes in *Hox* downstream target gene regulation without change in Hox expression, (3) changes in Hox protein function through changes in protein-coding sequence, and (4) posttranscriptional regulation of *Hox* gene function. Here, we focus on key examples from the literature during the past year that demonstrate each mechanism. For several of these case studies, direct links between *Hox* GRN changes and morphological evolution have been nicely demonstrated. For others, the challenge will be to determine the functional impact of *Hox* GRN changes.

Changes in *Hox* gene expression: examples of *cis*-regulatory evolution

Cis-regulatory changes in *Hox* GRNs have perhaps received the most attention in the literature, in keeping with the *cis*-regulatory hypothesis that postulates that genes involved in pattern formation and morphogenesis are highly constrained at the protein level but diversify because of changes in *cis*-regulatory elements.[3,55,56] This path of *cis*-regulatory evolution is thought to be favored because it increases flexibility while decreasing potentially negative consequences. Beacause most regulatory genes are pleiotropic—acting in different tissues and/or at different times during development—changes are permitted that alter expression of the regulatory gene in only specific body regions without affecting expression in other regions, thereby limiting the impact of such changes to only a subset of overall gene activity.[63,64] This *cis*-regulatory flexibility is explained in part by the modularity of *cis*-regulatory elements and the relative ease with which transcription factor binding sites can be gained and lost.[65–68] We discuss recent examples of *Hox* evolution because of *cis*-regulatory changes, including examples of dynamic changes in a rapidly evolving *Hox* gene, small variations in *Hox* expression domains, and acquisition of novel *Hox* expression patterns (Fig. 2, upper left panel).

The Hox gene fushi tarazu
Ancestrally, *fushi tarazu* (*ftz*) was likely expressed as a typical *Hox* gene, colinearly with its neighbors in the *Hox* complex (Figs. 1 and 3).[51] This *Hox*-like pattern is retained in extant arthropod species, including chelicerates (mite),[51] myriapods (millipede and centipede),[69,70] and a crustacean (water flea).[71] Yet, in *Drosophila*, *ftz* is not expressed in a *Hox*-like

pattern. Rather, it is expressed in a pair-rule pattern of seven stripes in the primordia of the alternate segmental regions missing in *ftz* mutants.[72–78] Expression of *ftz* in stripes is crucial for its pair-rule function: loss of stripe expression or ectopic expression of *ftz* outside the stripe domain each lead to lethality.[79] The dramatic change in *ftz* expression pattern from *Hox*-like to stripes was thought to have occurred in a basal insect lineage because striped *ftz* expression was observed in the firebrat *Thermobia* (Fig. 3, "+ stripes").[80] However, striped expression was not observed in the grasshopper, *Schistocerca* (Fig. 3, "– stripes").[81] Thus, either striped expression was lost in an Orthopteran lineage, or stripes were gained independently in basal insects (firebrat) and holometabolous insects (beetle, honeybee, and fruit fly), where all *ftz* genes examined are expressed in stripes (Fig. 3).[82,83]

Surprisingly, an additional change in *ftz* expression was recently observed in a crustacean, a phylogenetic group more closely related to insects than chelicerates or myriapods.[84] In the brine shrimp (*Artemia*), *ftz Hox*-like expression was virtually lost without any gain of an alternate mode of segmental expression. Other regulatory genes, such as the *Hox* gene *Antp* and segmentation gene *engrailed*, were expressed in typical *Hox*-like or segmentation patterns, respectively. However, *ftz* was only weakly expressed in a *Hox*-like pattern that was marginally detectable by *in situ* hybridization. Although *ftz* expression patterns from additional species in the Crustacea/basal Insecta groups are necessary to determine the order of changes in expression, these results suggest that *ftz Hox*-like expression was lost before pair-rule stripe expression was gained (Fig. 3). As discussed later, we suggest that this escape from colinearity was an enabling change that permitted further variation in *ftz* expression and function.

Scr expression varies among different insects
Several recent studies highlight the evolutionary flexibility of the *Hox* gene *Scr* (Fig. 3; see also section "cis-Regulatory changes in Hox target genes or novel protein functions"). In *Drosophila*, *Scr* (*Dm-Scr*) is expressed in the T1 segment primordial, and plays a role in patterning the labial appendages and the T1 segment, the latter role including suppression of wing development on this segment.[19,20,85,86] *Dm-Scr* also cooperates with other *Hox* genes to influence the formation of a dorsal ridge that

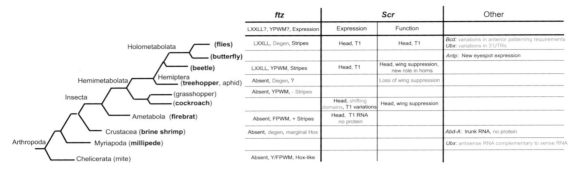

Figure 3. Cladogram summarizing the changes in *Hox* genes reported over the past year in arthropods. All species in the cladogram in bold are from recent studies. *Ftz* sequence and expression changes are listed as follows: LXXLL motif in green; degenerated YPWM motif in red; gain of stripes: "+ stripes," loss of stripes: "− stripes." For *Scr* and other *Hox* genes, changes that resulted in a gain of function are in green, and changes resulting in a loss of function are in red. Variations where the ancestral state is not known are in blue.

demarcates a tagmatic boundary between the insect head and thorax.[87,88] Previous studies of *Scr* expression and function were carried out in the beetle *Tribolium* (*Tc-Scr*) and the milkweed bug *Oncopeltus* (*Of-Scr*). *Tc-Scr* displayed similar functions to *Dm-Scr* in head development, T1 patterning, and in dorsal ridge formation; *Tc-Scr* expression in the beetle head is also similar to that seen in *Drosophila*.[89–91] Similarly, in the milkweed bug, *Scr* is required for head development and also plays a role in patterning T1, including the ability to suppress wing development.[92,93] Popadic's group has extended these earlier studies by examining *Scr* expression patterns in six ametabolous and hemimetabolous insects.[94] In all hemimetabolous species, *Scr* protein was found to accumulate in the head; however, variations in *Scr* expression were observed between species. These included shifts in the domains within the head that *Scr* was detected and variability as to whether and/or where *Scr* was expressed within the developing T1 leg primordia. Interestingly, in the basal insect *Thermobia*, no *Scr* protein was detected in T1, although *Scr* RNA expression was found (see also section "Posttranscriptional regulation of Hox genes").[94,95] The functional consequences of these variations in expression were examined using RNAi to knockdown *Scr* expression in the cockroach, *Periplaneta*.[96] *Scr* was found to be required for proper development of the labial palps, as seen in other insects. RNAi knockdown also resulted in an ectopic supernumerary segment between the head and first thoracic segment; this phenotype is similar to that observed in *Tribolium*.[91] Late RNAi effects revealed *Scr*'s role in wing suppression, as seen also in other insects including

hemimetabolous insects such as the milkweed bug[93] and treehoppers (see later),[97] and in holometabolous insects such as horned beetles (see later)[98] and fruit flies.[86] Interestingly, and in contrast to *Drosophila*, in neither the cockroach nor horned beetle did *Scr* RNAi affect the external morphology of T1 legs.[96,98] This observation led the authors to suggest that expression in T1 primordia preceded the function of *Scr* in T1-leg identity specification, as *Scr* is expressed in T1 in the cockroach. An alternate possibility that remains to be investigated is that leg identity was altered in more subtle ways that were not assessed in these experiments.

Together, these studies showed that both expression and function of *Scr* is highly conserved, although not identical, across these taxa. These studies arguably represent the most comprehensive comparative analysis of any arthropod *Hox* expression pattern; given that they revealed unexpected and subtle variability in expression, it is likely that many, if not all, *Hox* genes are undergoing similar small but significant changes. This is consistent with the suggestion of Angelini *et al.* that "evolution of insect *Hox* genes has included many small changes with general conservation of expression and function."[99] Future experiments will be important to reveal the relationships between the evolutionary variation in *Hox* expression patterns and their impact on function and morphology.

New expression domains suggest novel Antp functions in butterflies
A striking example of a gain of a novel *Hox* expression mode that is correlated with an evolutionary novelty was reported this year for butterfly

eyespots.[100] In the nymphalid butterfly *Bicyclus anynana*, a new expression pattern of *Antp* was observed. While still retaining its ancestral *Hox*-like expression pattern, *Antp* was also found to be expressed in a new domain in the organizing center of the eyespots. Previously, several highly conserved developmental genes, such as *Distalles (Dll)* and *engrailed*, were shown to be coopted for eyespot specification in butterflies.[101,102] Interestingly, *Antp* expression in the eyespot organizer region is earlier than these other regulatory genes, suggesting that it may play a critical role in initiating eyespot formation. This novel *Antp* expression pattern was also seen in several species closely related to *Bicyclus*, but was not found in *Junonia coenia*, a species with morphologically similar eyespots that diverged from *Bicyclus* approximately 90 million years ago. Future work will be needed to uncover the mechanisms that led to activation of *Antp* in this new expression domain in a certain lineages and to test the hypothesis that *Antp* indeed functions as an eyespot regulator, thereby linking the new expression pattern to morphological diversification.

cis-*Regulatory changes in vertebrate* Hox *complexes*

In a landmark study some years ago, Capecchi's group showed that the coding regions of *Hox* paralogs were functionally interchangeable in mice, thereby demonstrating that *cis*-regulatory change played a dominant role in the diversification of the *Hox* genes present in different clusters in vertebrates.[103] Recent work has extended these studies to elucidate the underlying evolutionary mechanisms. In one example, using a novel approach in which a targeted translocation was induced in the mouse genome,[104] Duboule's group placed the *HoxC* gene cluster under the regulatory control of the *HoxD* genomic locus and tested its ability to rescue *HoxD* loss-of-function phenotypes, which include defects in digit formation.[105] Their studies showed the *HoxC* cluster was largely able to rescue *HoxD* mutants, providing a compelling example of the importance of regulatory evolution within *Hox* complexes. Thus, after the duplication of *Hox* complexes in vertebrates, redundancy permitted diversification of highly related paralogs. This diversification seems to have occurred primarily at the level of *cis*-regulatory change, with the *Hox*

proteins themselves retaining ancestral and shared properties.

Differences in *Hox* gene regulation do not only apply across *Hox* complexes within a given species, but also are thought to be responsible for morphological differences between species. An important new study demonstrated that variations in *Hox* expression between birds and mammals in sensory systems that detect pain, touch, and other external stimuli result from differences in expression of *HoxD1*.[106] In mice, but not in chicks, the growth factor NGF induces expression of *HoxD1*. Mice lacking *HoxD1* develop altered neuronal circuitry that resembles that seen in chicks. Conversely, misexpression of *HoxD1* in the chick induced an axonal patterning similar to that seen in the mouse. These studies thus revealed a novel role for a *HoxD1* GRN in wiring of the sensory system in vertebrates. Importantly, they implicate a change in *HoxD1* expression, and define its origin—a switch in responsiveness to growth factor signaling—as the causal switch in an important functional difference between species.

The role of *Hox* genes in patterning the axial skeleton was found early on, suggesting that modifications of the *Hox* code would have important ramifications for evolutionary diversification of skeletal structures.[107,108] Two studies this year support this idea. DiPoi *et al.* expanded upon the work of Cohn *et al.*[109] who suggested that expansion of *Hox* gene expression domains in snakes accounts for the morphological novelties seen in this taxa. Comparative genomic and expression analysis of *Hox* genes in various groups of reptiles revealed conservation and divergence in both the expression and function of these genes.[110] Comparing an extant snake and lizard, the authors showed that changes in *Hox10* and *Hox13* expression correlate with alterations in the snake axial skeleton, which has expanded relative to the lizard and the presumed ancestral state. They further speculated that snakes use a simplified *Hox* code, due in part to changes in *Hox* expression, which may have driven the innovations in the snake axial skeleton. In a related comparative study, the expression of *Hox* gene paralogs (groups 4–8) that pattern cervical and thoracic regions of the mammalian axial skeleton were compared between birds and alligators, the latter representing the closest extant reptilian relative to birds (∼250 million years of divergence). Changes in the expression domains of two *Hox* genes were noted: first, loss of a portion of *Hoxc8* expression correlates with unique morphological features in the alligator thorax and, second, patterns of *Hoxa5* correlate with differential cervical structures in chick, alligator, as well as mammals.[111] Together, these studies provide nice examples of how changes in the expression of *Hox* genes may have driven the evolution of novel morphologies.

cis-Regulatory changes in *Hox* target genes or novel protein functions

In some evolutionary scenarios, new biological functions of a *Hox* gene have been observed without corresponding changes in the expression pattern of that *Hox* gene. In such cases, the change in phenotype may be a result of changes in the *cis*-regulatory regions of downstream target genes[3] (Fig. 2, upper right panel) or changes in the *Hox* protein that alter its regulatory specificity (see "Changes in *Hox* protein function" below, Fig. 2 lower left panel). For all of the examples discussed in this section, future studies are required to distinguish between these mechanisms.

Scr *is a key player in insect morphological evolution*

Two striking examples of the genetic basis of morphological evolution both result from novel functions of the *Hox* gene *Scr* (Fig. 3). The first example links *Scr* to the horns of dung beetles, which differ dramatically in size and shape (Fig. 4A–C). These horns, which develop in the prepupal stage as epidermal outgrowths of the head and/or prothorax and then undergo remodeling during the pupal stage, are diverse and dramatic in appearance in adults.[112,113] Their location suggested that the *Hox* gene *Scr* might be involved in their patterning. To examine this, *Scr* orthologs were isolated from two horned beetle species, and expression and function were assessed by RNAi knock-down.[98] The expression pattern of *Scr* was *Hox*-like: both mRNA and *Scr* protein were expressed in patterns similar to those seen in other insects. Interestingly, RNAi experiments revealed a novel role for *Scr* in growth of the pronotal horns. These effects differed somewhat between the two species, which also showed sex differences in response to *Scr* RNAi, suggesting variations in *Scr* function and its interaction with sex determination pathways across species. To-

gether, these results demonstrate that a morphological novelty in this group of dung beetles, which is not seen in the numerous other insect lineages that express *Scr* in the same domain, results from changes in *Scr* function. *Scr* was coopted into at least one new developmental pathway, allowing it to acquire a new function with distinct effects on morphology without noticeable change in its typical *Hox*-like expression pattern or loss of its "traditional" *Hox* roles in body patterning. It will be of great interest to delve deeper into the mechanism underlying *Scr* gain-of-function in horn development: Did changes in *cis*-regulatory regions bring new sets of genes under *Scr* control? Did changes in *Scr* function—in the expression of an *Scr* cofactor, or the ability of *Scr* to interact with transcriptional activators or repressors—alter *Scr*-regulatory specificity such that a new set of genes came under its control?

A second elegant example of changing *Scr* function without change in its expression pattern was found for treehopper helmets.[97] Treehoppers are a large group of diverse hemipteran insects that share a novel helmet structure, which manifests in a remarkable array of appearances (Fig. 4D–F). Similar to the logic for beetle horns discussed above, the location of the helmet suggested a potential role for *Scr*. In their recent study, Prud'homme *et al.* showed that *Scr* expression in treehoppers is similar to that in other insects (*Hox*-like), but helmets, novel wing-like structures, have been allowed to develop and diversify on T1 because *Scr* has lost the ability to repress genes necessary for wing development. One of these genes is *nubbin* (*nub*): *nub* is necessary for wing development in *Drosophila* and is absent from T1.[114] However, in the treehopper *Publilia modesta*, *nub* was detected in the developing helmet in a pattern similar to that of developing wings. This suggests that *Pm-Scr* has lost the ability to downregulate *nub* expression, as well as other wing specification genes, and this was likely a critical step in the morphological evolution of this structure. The authors further showed that ectopic expression of treehopper *Scr* repressed wing formation in *Drosophila*, suggesting that change in the function of *Scr* does not explain its loss of ability to repress wing-realizator genes. As discussed in the previous section, these results leave open two possible mechanistic explanations: first, that changes in the *cis*-regulatory regions of target genes

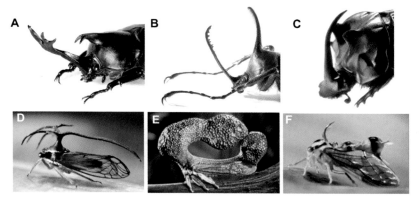

Figure 4. Diversity in beetle horn and treehopper helmet morphologies. Changes in *Scr* function but not expression have resulted in novel morphologies on the T1 segment of beetles and treehoppers. (A–C) Photographs showing the diversity in horns of dung beetles. (D–F) Photographs showing treehopper helmet diversity, which were found to be wing-like appendages on the T1 segment.[97] Figure modified with permission from Moczek.[113]

rendered them unresponsive to *Scr* repression; second, that changes in an *Scr*-cofactor interaction altered its regulatory specificity such that wing-realizator genes were no longer negatively regulated. Although the former is likely (*cis*-regulatory changes in *Hox* targets), further experiments are required to distinguish between these mechanisms (e.g., see approach of Refs. 115–117). The incredible morphological diversity of treehopper helmets and the rapid progress in identifying the patterning genes controlling its development make this an exciting system for working out molecular mechanisms leading to development and differentiation of complex and evolutionarily plastic body structures.

Changes in *Hox* protein function

Hox genes are generally considered to be highly conserved and evolutionarily constrained at the level of protein activity. This conclusion comes largely from transspecies experiments in which *Hox* genes from distant taxa were expressed in *Drosophila* and demonstrated conserved function (see section "Introduction").[38–42] However, evidence is continuing to emerge that changes in Hox protein sequence occur and that these changes can lead to changes in the functional properties of Hox proteins by altering their regulatory specificity (Fig. 2, lower left panel).[58,118–120] An important example of this type of change was provided several years ago by groups studying the role of *Hox* genes in patterning abdominal appendages. In contrast to insects, crustaceans have appendages on posterior segments.[121,122] The

limbless insect abdomen is thought to be explained in part by the ability of Ubx to repress the target gene *Dll*.[123] In contrast to *Dm*-Ubx, Ubx proteins from an onychophoran and a crustacean did not repress *Dll* when expressed in *Drosophila*. It was further shown that insect Ubx proteins have taken on a role in *Dll* repression by the acquisition of a repressor domain, which is missing or nonfunctional in noninsect Ubx proteins. This suggests that the acquisition of a repressor function, because of specific changes in Ubx protein sequences in insect lineages, contributed to the evolution of the limbless abdomen in insects. Below we discuss examples from the last year that have implicated motifs in other Hox proteins with changes in function.

Escape from colinearity enabled variation in Hox protein potential

Ftz evolution and the importance of cofactor-interaction motifs. Although *ftz* is in the *Hox* complex, the gene was first characterized in *Drosophila* as a pair-rule segmentation gene.[72,74] As discussed above, the change in *ftz* function from an ancestral homeotic gene to a pair-rule gene in *Drosophila* is explained in part by its change in expression. However, the distinctly nonhomeotic function of *ftz* in *Drosophila* led us to ask several years ago whether changes had also occurred in *Drosophila* Ftz protein (*Dm*-Ftz) to switch its function from a homeotic to a pair-rule segmentation protein.[124] These studies showed that *Dm*-Ftz has lost homeotic potential: even when misexpressed in

Drosophila in a homeotic fashion, *Dm*-Ftz does not produce homeotic transformations. In contrast, Ftz proteins from the grasshopper and beetle did produce homeotic transformations when misexpressed in *Drosophila,* and they also regulated *Antp* target genes indicative of conserved *Antp*-like homeotic function. These differences in protein function were attributed at least in part to two changes in cofactor interaction motifs in Ftz proteins during arthropod evolution. First, *Dm*-Ftz acquired an LXXLL motif that is necessary for its functional interaction with a novel cofactor, the orphan nuclear receptor Ftz-F1. Second, the YPWM motif, which mediates interaction with the *Hox* cofactor Extradenticle (Exd) has degenerated in *Dm*-Ftz. Ftz from the beetle, which displayed both homeotic and segmentation function when expressed in *Drosophila,* has both LXXLL and YPWM motifs. Swaps of these motifs demonstrated that addition of a YPWM to *Dm*-Ftz rendered it capable of causing homeotic transformation, while simultaneous inactivation of the LXXLL motif, along with addition of YPWM, increased the homeotic potential of this protein.[125]

These studies led to the hypothesis that *Dm-ftz* acquired an exclusive role in segmentation because of at least two major functional changes in protein sequence: (1) Ftz protein acquired an LXXLL motif making possible a functional interaction with cofactor Ftz-F1 and (2) its ancestral YPWM motif degenerated, diminishing interaction with Exd and decreasing its ability to regulate homeotic gene targets. In a recent study, we tracked these changes through arthropod phylogeny to determine when and in what order they occurred (Fig. 3).[84,126] First, we found that the LXXLL motif was stably acquired at the base of the holometabolous insects (Fig. 3, green LXXLL). Ftz sequences from every holometabolous insect available to date (12 *drosophilid* species and 7 other holometabolous insects, spanning ~300 million years of evolutionary divergence) retain an LXXLL sequence at their N-termini and this sequence is most often LRALL (17/19 *ftz* sequences). The retention of this sequence through the holometabolous lineage is highly suggestive of an adaptive gain-of-function change in coding sequence. Second, and in contrast to the LXXLL sequence, the YPWM motif degenerated multiple times during the radiation of the arthropods (Fig. 3, red "degen"). The degenerate YPWM motifs differ in sequence in different lineages, with

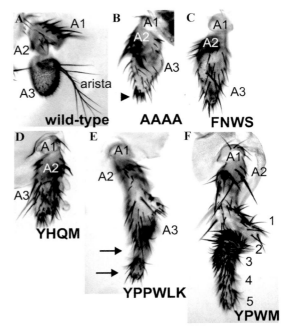

Figure 5. Degenerate YPWM motifs retain varying degrees of homeotic potential. The *ftz* transgenes carrying examples of natural variation in YPWM motifs were expressed in developing imaginal discs with the *Dll-Gal4* driver. (A) Control, expression of *UAS-lacZ* did not cause homeotic transformation of antennae. (B) *Dm-ftz*-AAAA animals showed normal A1 and A2, but abnormal A3 segments with bristles (arrowhead) and no aristae. (C) *Dm-ftz*-FNWS and (D) *Dm-ftz*-YHQM effects were similar to *Dm-ftz*-AAAA. (E) *Dm-ftz*-YPPWLK caused transformation of aristae into partial legs with two segments (arrows) and malformed A3 segments. (F) *Dm-ftz*-YPWM caused complete transformation of aristae to legs with five segments. Figure used with permission from Heffer *et al.*[84]

some even having insertions in the motif (e.g., IPQM, YHQM, and YPPWLK). Given the repeated degeneration of the YPWM motif, we next asked if homeotic potential conferred by this motif was undergoing functional loss as this motif degenerated. By inserting the degenerate motifs into the *Dm-ftz* coding sequence, we showed that the YPWM motif itself conferred homeotic function while different degenerate motifs conferred varying degrees of homeotic potential, suggesting that they are in the process of losing their homeotic potential (Fig. 5).

How were such changes tolerated, and even possibly adaptive, when equivalent changes (forced gain or loss of expression or misexpression of altered proteins) made in the laboratory are almost always detrimental? Although more work needs to be done to address this question, we proposed that it was the initial loss-of-*Hox*-like expression of *ftz* that freed the protein from functional constraints, allowing escape from colinearity. Our model proposes that loss of expression was itself an enabling change, allowing for alterations in protein sequence that would otherwise have been detrimental to embryonic development; expression at low levels was tolerated, allowing a protein with novel specificity to "hover beneath the surface" until it was later coopted into a new developmental pathway. This situation is somewhat reminiscent of that seen in a long-term evolution experiment in bacteria, where it was shown that mutations that promote the long-term fitness of a population do not confer an immediate survival advantage. Rather, these initially less-fit bacteria prevailed in the population because of their increased ability to adapt.[127] We speculate that, similarly, a regulatory gene that lost its homeotic function did not provide an immediate fitness advantage, but rather that this gene was made available for later cooption into new developmental pathways, a situation that ultimately proved to be of some—yet to be determined—benefit. Many questions remain to be answered here. For example, what was the impact of the gain of a striped expression pattern on embryonic development? Do Ftz proteins that have degenerate YPWM motifs play roles in homeosis in the organisms from which they were derived? Was the ancestral role of the LXXLL an interaction with Ftz-F1 and a role in segmentation, or did the LXXLL stabilize earlier because of some other as yet unknown function, followed by cooption into segmentation pathways? Although we can hypothesize based on current data, answering these questions will require functional experiments in the species that express varying forms of *ftz*.

Hox3/zen duplication and divergence: Bcd acquires a new function in higher insects. Similar to *ftz*, insect *zen* genes have escaped the constraints of colinearity and diverged dramatically in expression and protein function. In higher Diptera, a duplication of *zen* produced *bcd*, a novel *Hox* gene that took on a unique role in anterior patterning because of its expression at the embryonic anterior pole and unique modifications of its protein sequence, including a novel DNA-binding specificity and the ability to bind to RNA and thus regulate transla-

tion.[57,128–131] As *bcd* has been reviewed extensively, we mention here only one interesting new finding from this year examining the sequence, expression, and function of the *bcd* gene from a "lower" fly.[132] *Episyrphus bicoid* (*Eb-bcd*) is localized to the anterior pole of the embryo, as is *Dm-bcd*. However, although *Eb-bcd* protein is similar to *Dm-bcd*, being a clear ortholog with a similar homeodomain, it lacks several of the sequence motifs that are important for *Dm-bcd* function, suggesting differences in the biochemical properties of the two proteins. Further, RNAi experiments showed that *Eb-bcd* is the major anterior determinant in this fly and that it is responsible for additional aspects of patterning, such as regulation of gap gene expression. This latter role of *Eb-bcd* function is shared among several different maternal gene products in *Drosophila*. Thus, variations in the protein sequence and biological functions of *bcd* were observed, despite a shared expression pattern. Future experiments will determine whether *Eb*-Bcd gained additional functions, which are carried out by different genes in *Drosophila*, or *Dm*-Bcd lost some of the regulatory potential of a shared Bcd ancestor.

Hox *protein changes and the evolution of placental mammals*

Perhaps the best-documented example of a change in *Hox* protein sequence implicated in the evolution of a morphological novelty is mammalian *HoxA11*, which underwent a period of adaptive evolution in the stem lineage of placental mammals to take on roles in the establishment and maintenance of pregnancy.[133] It was previously shown that *HoxA11* is a transcriptional activator of prolactin, a gene critical for establishment of pregnancy in mammals. *HoxA11* from placental mammals activated prolactin gene expression, while *HoxA11* from birds or nonplacental mammals (opossum, platypus) did not. This functional difference in *HoxA11* was explained by changes in HoxA11 protein that allowed for interaction with a new partner, *Foxo1a*.[134,135] In a study this year, Wagner's group tested the biochemical basis of this species-specific protein–protein interaction. Coimmunoprecipitation experiments were carried out with proteins from extant species as well as reconstructed ancestral HoxA11 and Foxo1a proteins.[136] *Foxo1a* interacted with HoxA11 from placental mammals (human, opossum) and with ancestral eutherian, therian, and

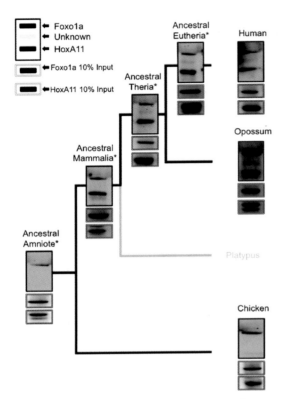

Figure 6. Physical interaction between HoxA11 and Foxo1a originated in the mammalian stem lineage. Note the presence of a band for human, opossum, ancestral eutherian, ancestral therian, and ancestral mammalian HoxA11-V5/His, but there is no band for chicken or ancestral amniote HoxA11-V5/His, indicating that the physical interaction arose in the mammalian stem lineage. The legend in the upper right corner of the figure indicates location/identity of bands shown in panels, including unknown peptides (observed in some panels) that cross-react with the antibody. Asterisks (*) indicate reconstructed ancestral proteins. Flag epitope–tagged *Foxo1a* and V5/His epitope–tagged *HoxA11* mammalian expression vectors were cotransfected into HeLa cells, and nuclear lysates were incubated with anti-Flag agarose overnight. The following day, samples were treated with DNaseI and washed, and protein complexes were eluted with NuPage LDS sample buffer. Protein complexes were resolved by SDS/PAGE and transferred to a PDVF membrane. Blots were probed with anti-Flag and anti-V5 antibodies. Experiments were repeated a minimum of three times, and the results presented here are representative of typical findings. Figure reprinted with permission from Brayer *et al.*[136]

mammalian HoxA11. However, Foxo1a failed to interact with extant bird (chicken) HoxA11 (Fig. 6). To determine whether changes in HoxA11, Foxo1a, or both, facilitated the acquisition of this protein–protein interaction, binding of human HoxA11 to a range of Foxo1a proteins was examined; all Foxo1a

proteins interacted with human HoxA11, showing that changes in Foxo1a were not necessary for the functional switch. Rather, changes in HoxA11 protein permitted a new interaction with Foxo1a. Thus, interaction between HoxA11 and Foxo1a only occurs in mammals, despite the fact that both proteins are present in out-groups. Interestingly, this interaction arose in a mammalian stem lineage, before HoxA11/Foxo1a acquired the ability to regulate prolactin gene expression—a feature that arose later, in placental mammals. It will be of great interest to see what the original role was of the HoxA11/Foxoa1 pair, before its recruitment for prolactin regulation, what the factors were that allowed and selected for this gain-of-function change in placental mammals, and to what extent this single switch in function of a transcriptional regulator explains the emergence of placental development. Irrespective of the outcome of these future endeavors, the careful biochemical approaches taken in this work bring mechanistic analysis of *Hox* evolution to a new level.

Hox *protein evolution in reptiles*

As mentioned above, *Hox* evolution in reptiles has occurred in part through changes in *cis*-regulatory regions. In addition, residues in *Hox* protein sequence were found to be under selection in Hox10 paralogs; specifically, a comparison of the first exon of snake *Hoxa10, -c10,* and *–d10* genes identified three motifs where purifying selection was relaxed.[110] Protein changes were also observed in reptilian *Hox11* orthologs. In 3′UTR snakes, colinear expression is retained, but rib-suppressing functions seem to have been lost. Here, it was proposed that positive selection on specific sets of amino acids decreased the rib-suppressing activity of HoxA11.[110] Although it may be coincidental, it is interesting to note that changes in HoxA11 function have contributed to two very different evolutionary innovations: the appearance of placental mammals and major changes in the reptilian axial skeleton.

In sum, changes in Hox protein sequence have been observed in many cases to date, and some of these changes have been correlated with morphological evolution. These findings are surprising because early studies of Hox proteins suggested that their function was highly conserved across very divergent taxa, such as human and fly. The strong similarity between homeodomain sequences across taxa also supported this idea of functional constraint, although clear exceptions to this exist, for example the case of Bcd.[128] However, early comparative studies largely ignored the regions of Hox proteins outside of the homeodomain, which have often diverged so rapidly that they show no sequence similarities. These portions of Hox proteins are not dispensable, as sometimes assumed; rather, they provide a rich source of functional potential for the forces of evolution to act upon. The accumulating evidence for functional changes in Hox proteins, some of which was reviewed here, challenges the view that Hox protein change is a rare exception and suggests that change in the function of regulatory genes plays a significant role in body plan evolution.

Posttranscriptional regulation of *Hox* genes

Hox genes may also be regulated posttranscriptionally such that the expression and function of Hox proteins is modulated without affecting *cis*-regulatory or coding sequences (Fig. 2, lower right panel). This regulatory mechanism has not received much attention in the past, but is highlighted by new findings this year.

Several examples were reported recently in which Hox protein is not detectable in regions of embryos where mRNA is found (Fig. 3). In a crustacean, the brine shrimp *Artemia*, *Abd-A* mRNA was detected in a *Hox*-like pattern in the trunk region of early embryos, but Abd-A protein was undetectable.[137] When the shrimp protein, *Af*-Abd-A, was expressed in *Drosophila*, no protein was detected either. Together, this suggests a change in *Abd-A* mRNA that decreased its stability or translation efficacy, or both. The absence of Abd-A protein in the trunk likely contributes to the ability of this species to develop limbs throughout the trunk region—a phenomenon that is repressed by both Ubx and Abd-A in insects that lack abdominal legs.[137] Similarly, discrepancies between the mRNA and protein patterns were reported for *Scr* in both *Thermobia* and *Oncopeltus* (Fig. 3).[94,95]

Additional studies this year provide evidence for regulation of *Hox* genes at the mRNA level. Studies from the Alonso lab showed that differential 3′UTR formation in *Ubx* generates targets for regulation by different miRNAs and that this occurs in a developmental and tissue-specific fashion.[138] Building upon this, Patraquim *et al.* compared sequences of

Hox gene 3′UTR sequences from the 12 sequenced *drosophilid* genomes. They found that these 3′UTRs are evolving (as would be expected for any nucleotide sequence), but while the sequences differ greatly, the topology of these regions seems to be under strong selective pressure, suggestive of functional constraint. The changes seen in 3′UTR sequence include changes in potential regulatory sites for miRNAs.[139] Finally, in a recent study in millipedes, a *Ubx* antisense transcript was found that is expressed in a pattern complementary to *Ubx*-coding RNA, suggesting that the antisense RNA is negatively regulating the transcription or stability of *Ubx* sense RNA.[140] Although the mechanism remains to be determined, this scenario is reminiscent of the noncoding RNAs at the *bithoraxoid* (*bxd*) locus in *Drosophila*, which repress *Ubx* expression in *cis* by transcriptional interference.[141]

Taken together, these findings document additional levels at which evolution tinkers with *Hox* function. In this context, observations from the Hogness lab in the early 1980s may have been visionary, as they suggested, long before RNAi was a common tool of molecular geneticists, that "The elements of the *bxd* region might make RNA products that interact with the *Ubx* RNA or with other small RNA molecules involved in processing *Ubx* RNA."[142,143]

Conclusions and emerging themes

This year has seen many new examples of evolutionary flexibility in *Hox* genes, despite the fact that these genes were thought to be highly constrained because of their essential roles in embryonic patterning. These genes, which were once thought to be highly static building blocks of the animal body plan, are in fact changing, and change is occurring at many levels. In the examples reviewed here, both gain and loss of *Hox* activity has been observed, and no clear pattern has emerged as to which is more frequent. In some cases, *Hox* genes have been coopted into new developmental pathways during evolution without loss of "traditional" *Hox*-like functions (e.g., cooption of *Scr* into regulation of beetle horns or treehopper helmets). In others, redundancy of duplicate genes has allowed for subfunctionalization or neofunctionalization (e.g., in the case of *Antp* and *ftz*, *Antp* has maintained the traditional *Hox*-like roles, freeing-up *ftz* to diverge). Still, in other cases, new expression domains have emerged, because of *cis*-regulatory changes in the *Hox* genes or changes in upstream activators (e.g., expression of *Antp* in the eyespot primordia of butterflies or novel regulation of *Hoxd1* by NGF in neuronal patterning). In some cases, variations in *Hox* expression patterns have yet to be correlated with specific morphological divergence, but their variation, while in some ways subtle, is much more extensive than previously imagined (e.g., *Scr* expression within the head and thorax of diverse insects). Finally, the importance of posttranscriptional control of *Hox* function has emerged, making extrapolation about function from *in situ* hybridization data even more difficult.

In sum, in this year alone, many new examples of evolutionary flexibility in *Hox* genes have been uncovered. These changes occur at all four mechanistic levels discussed here (Fig. 2). Much remains to be learned about these and other as yet undiscovered mechanisms underlying *Hox* variation and the functional consequences on development and evolution. In our view, change will not be limited to one type of mechanism and, although one may be more frequent than another, the diversification of the animal body plan is a matter too rich in complexity to be boiled down to one simple explanation. With new animal systems and experimental approaches arising to address these issues, we expect rapid progress in our understanding of the impacts of evolutionary tinkering with *Hox* genes in coming years.

Acknowledgment

We thank Jeff Shultz for helpful discussions and the NSF (IOS0641717) for support.

Conflicts of interest

The authors declare no conflicts of interest.

References

1. Erwin, D.H. & E.H. Davidson. 2002. The last common bilaterian ancestor. *Development* **129:** 3021–3032.
2. Gilbert, S.F. 2003. *Developmental Biology.* Sinauer Associates. Sunderland, MA.
3. Carroll, S.B., J.K. Grenier & S.D. Weatherbee. 2005. *From DNA to Diversity: Molecular Genetics and the Evolution of Animal Design.* 2nd ed. Blackwell Science Ltd. Malden, MA.
4. Bateson, W. 1984. *Materials for the Study of Variation Treated with Especial Regards to Discontinuity in the Origin of Species.* Macmillan. London.
5. Lewis, E.B. 1978. A gene complex controlling segmentation in *Drosophila. Nature* **276:** 565–570.
6. Lewis, E.B. 1998. The bithorax complex: the first fifty years. *Int. J. Dev. Biol.* **42:** 403–415.
7. Duncan, I. 1987. The bithorax complex. *Annu. Rev. Genet.* **21:** 285–319, doi: 10.1146/annurev.ge.21.120187.001441.
8. Gehring, W.J. 1966. Bildung eines vollstandigen Mittlebeines mit Sternopleura in der Antennenregion ei der Mutante *Nasobemia (Ns)* von *Drosophila melanogaster. Arch. Jul. Klaus Stift. Vererbungsforsch.* **41:** 44–54.
9. Postlethwait, J.H. & H.A. Schneiderman. 1971. Pattern formation and determination in the antenna of the homoeotic mutant Antennapedia of *Drosophila melanogaster. Dev. Biol.* **25:** 606–640.
10. Denell, R.E. 1973. Homoeosis in Drosophila. I. Complementation studies with revertants of Nasobemia. *Genetics* **75:** 279–297.
11. Duncan, I.W. & T.C. Kaufman. 1975. Cytogenic analysis of chromosome 3 in *Drosophila melanogaster:* mapping of the proximal portion of the right arm. *Genetics* **80:** 733–752.
12. Kaufman, T.C., R.A. Lewis & B.T. Wakimoto. 1980. Cytogenetic analysis of chromosome 3 in *Drosophila melanogaster.* The homeotic gene complex in polytene chromosome interval 84A-B. *Genetics* **94:** 115–133.
13. Lewis, R.A., B.T. Wakimoto, R.E. Denell & T.C. Kaufman. 1980. Genetic analysis of the Antennapedia gene complex (ANT-C) and adjacent chromosomal regions of *Drosophila melanogaster.* II. Polytene chromosome segments 84A-84B1,2. *Genetics* **95:** 383–397.
14. Denell, R.E., K.R. Hummels, B.T. Wakimoto & T.C. Kaufman. 1981. Developmental studies of lethality associated with the Antennapedia gene complex in *Drosophila melanogaster. Dev. Biol.* **81:** 43–50.
15. Schneuwly, S., A. Kuroiwa, P. Baumgartner & W.J. Gehring. 1986. Structural organization and sequence of the homeotic gene Antennapedia of *Drosophila melanogaster. EMBO J.* **5:** 733–739.
16. Schneuwly, S., A. Kuroiwa & W.J. Gehring. 1987. Molecular analysis of the dominant homeotic *Antennapedia* phenotype. *EMBO J.* **6:** 201–206.
17. Schneuwly, S., R. Klemenz & W.J. Gehring. 1987. Redesigning the body plan of Drosophila by ectopic expression of the homoeotic gene *Antennapedia. Nature* **325:** 816–818.
18. Struhl, G. 1982. Genes controlling segmental specification in the *Drosophila* thorax. *Proc. Natl. Acad. Sci. USA* **79:** 7380–7384.
19. Mahaffey, J.W. & T.C. Kaufman. 1987. Distribution of the *Sex combs reduced* gene products in *Drosophila melanogaster. Genetics* **117:** 51–60.
20. LeMotte, P. K., A. Kuroiwa, L.I. Fessler & W.J. Gehring. 1989. The homeotic gene *Sex combs reduced* of *Drosophila:* gene structure and embryonic expression. *EMBO J.* **8:** 219–227.
21. Lewis, E.B. 1997. In *Nobel Lectures, Physiology or Medicine 1991–1995.* Nils Ringertz, Ed. World Scientific Publishing Co. Singapore.
22. Laughon, A. & M.P. Scott. 1984. Sequence of a *Drosophila* segmentation gene: protein structure homology with DNA-binding proteins. **310:** 25–31.
23. Weiner, A.J., M.P. Scott & T.C. Kaufman. 1984. A molecular analysis of *fushi tarazu,* a gene in *Drosophila melanogaster* that encodes a product affecting embryonic segment number and cell fate. *Nature* **37:** 843–851.

24. McGinnis, W., R.L. Garber, J. Wirz, *et al.* 1984. A homologous protein-coding sequence in *Drosophila* homeotic genes and its conservation in other metazoans. *Cell* **37:** 403–408.

25. McGinnis, W., M.S. Levine, E. Hafen, *et al.* 1984. A conserved DNA sequence in homeotic genes of the *Drosophila* Antennapedia and bithorax complexes. *Nature* **308:** 428–433.

26. Desplan, C., J. Theis & P.H. O'Farrell. 1985. The *Drosophila* developmental gene, *engrailed*, encodes sequence-specific DNA binding activity. *Nature* **318:** 630–635.

27. Shepherd, J.C., W. McGinnis, A.E. Carrasco, *et al.* 1984. Fly and frog homoeo domains show homologies with yeast mating type regulatory proteins. *Nature* **310:** 70–71.

28. Gehring, W.J. 1985. The homeo box: a key to the understanding of development? *Cell* **40:** 3–5, doi: 0092-8674(85)90300-9 [pii].

29. Gehring, W.J. & Y. Hiromi. 1986. Homeotic genes and the homeobox. *Ann. Rev. Genet.* **20:** 147–173.

30. Lawrence, P.A. 1992. *The Making of a Fly: The Genetics of Animal Design.* Blackwell Scientific Publications. Oxford.

31. Gehring, W.J. *et al.* 1994. Homeodomain-DNA recognition. *Cell* **78:** 211–233.

32. McGinnis, W. 1994. A century of homeosis, a decade of homeoboxes. *Genetics* **137:** 607–611.

33. McGinnis, W. & R. Krumlauf. 1992. Homeobox genes and axial patterning. *Cell* **68:** 283–302.

34. Pearson, J. C., D. Lemons & W. McGinnis. 2005. Modulating Hox gene functions during animal body patterning. *Nat. Rev. Genet.* **6:** 893–904.

35. Garcia-Bellido, A. 1975. Genetic control of wing disk development in *Drosophila.* In Cell Patterning, *Ciba. Found. Symp.* **29:** 161–182.

36. McGinnis, W., C.P. Hart, W.J. Gehring & F.H. Ruddle. 1984. Molecular cloning and chromosome mapping of a mouse DNA sequence homologous to homeotic genes of Drosophila. *Cell* **38:** 675–680.

37. Lufkin, T. *et al.* 1992. Homeotic transformation of the occipital bones of the skull by ectopic expression of a homeobox gene. *Nature* **359:** 835–841.

38. Malicki, J., L.D. Bogarad, *et al.* 1993. Functional analysis of the mouse homeobox gene HoxB9 in *Drosophila* development. *Mech. Dev.* **42:** 139–150.

39. Malicki, J., K. Schughart & W. McGinnis. 1990. Mouse *Hox-2.2* specifies thoracic segmental identity in *Drosophila* embryos and larvae. *Cell* **63:** 961–967.

40. McGinnis, N., M.A. Kuziora & W. McGinnis. 1990. Human *Hox-4.2* and *Drosophila Deformed* encode similar regulatory specificities in *Drosophila* embryos and larvae. *Cell* **63:** 969–976.

41. Zhao, J.J., R.A. Lazzarini & L. Pick. 1993. The mouse *Hox-1.3* gene is functionally equivalent to the *Drosophila Sex combs reduced* gene. *Genes Dev.* **7:** 343–354.

42. Zhao, J.J., R.A. Lazzarini & L. Pick. 1996. Functional dissection of the mouse *Hox-a5* gene. *EMBO J.* **15:** 1313–1322.

43. Aubin, J., P. Chailler, D. Menard & L. Jeannotte. 1999. Loss of Hoxa5 gene function in mice perturbs intestinal maturation. *Am J Physiol* **277:** C965–C973.

44. Aubin, J., U. Dery, M. Lemieux, *et al.* 2002. Stomach regional specification requires Hoxa5-driven mesenchymal-epithelial signaling. *Development* **129:** 4075–4087.

45. Aubin, J., M. Lemieux, J. Moreau, *et al.* 2002. Cooperation of Hoxa5 and Pax1 genes during formation of the pectoral girdle. *Dev. Biol.* **244:** 96–113.

46. Garin, E., M. Lemieux, Y. Coulombe, *et al.* 2006. Stromal Hoxa5 function controls the growth and differentiation of mammary alveolar epithelium. *Dev. Dyn.* **235:** 1858–1871.

47. Jeannotte, L., M. Lemieux, J. Charron, *et al.* 1993. Specification of axial identity in the mouse: role of the *Hox-a5* (*Hox 1.3*) gene. *Genes Dev.* **7:** 2085–2096.

48. Mandeville, I. *et al.* 2006. Impact of the loss of Hoxa5 function on lung alveogenesis. *Am. J. Pathol.* **169:** 1312–1327.

49. Grenier, J.K., T.L. Garber, R. Warren, *et al.* 1997. Evolution of the entire arthropod Hox gene set predated the origin and radiation of the onychophoroan/arthropod clade. *Curr. Biol.* **7:** 547–553.

50. Cook, C.E., M.L. Smith, M.J. Telford, *et al.* 2001. Hox genes and the phylogeny of the arthropods. *Curr. Biol.* **11:** 759–763.

51. Telford, M.J. 2000. Evidence for the derivation of the Drosophila fushi tarazu gene from a Hox gene orthologous to lophotrochozoan Lox5. *Curr. Biol.* **10:** 349–352.

52. Wagner, G.P., C. Amemiya & F. Ruddle. 2003. Hox cluster duplications and the opportunity for evolutionary novelties. *Proc. Natl. Acad. Sci. USA* **100:** 14603–14606.

53. Scott, M.P. 1992. Vertebrate homeobox gene nomenclature. *Cell* **71:** 551–553.

54. Duboule, D. 1994. *Guidebook to the Homeobox Genes.* Oxford University Press. New York.

55. Prud'homme, B., N. Gompel & S.B. Carroll. 2007. Emerging principles of regulatory evolution. *Proc. Natl. Acad. Sci. USA* **104**(Suppl 1), 8605–8612, doi: 0700488104 [pii]10.1073/pnas.0700488104.

56. Carroll, S.B. 2008. Evo-devo and an expanding evolutionary synthesis: a genetic theory of morphological evolution. *Cell* **134:** 25–36.

57. Schmidt-Ott, U. & E.A. Wimmer. 2004. In *Modularity in Development and Evolution.* G.P. Wagner, Ed. The University of Chicago Press. London.

58. Lynch, V.J. & G.P. Wagner. 2008. Resurrecting the role of transcription factor change in developmental evolution. *Evolution* **62:** 2131–2154.

59. Heffer, A., U. Lohr & L. Pick. 2011. *ftz* evolution: findings, hypotheses and speculations (response to DOI 10.1002/bies.201100019). *Bioessays* **33:** 910–918.

60. Britten, R.J. & E.H. Davidson. 1969. Gene regulation for higher cells: a theory. *Science* **165:** 349–357.

61. Davidson, E.H. & D.H. Erwin. 2006. Gene regulatory networks and the evolution of animal body plans. *Science* **311:** 796–800.

62. Levine, M. & E.H. Davidson. 2005. Gene regulatory networks for development. *Proc. Natl. Acad. Sci. USA* **102:** 4936–4942, doi: 10.1073/pnas.0408031102.

63. Stern, D.L. & V. Orgogozo. 2008. The loci of evolution: how predictable is genetic evolution? *Evolution* **62:** 2155–2177.

64. Stern, D.L. & Orgogozo, V. 2009. Is genetic evolution predictable? *Science* **323:** 746–751.

65. Howard, M.L. & E.H. Davidson. 2004. cis-Regulatory control circuits in development. *Dev. Biol.* **271:** 109–118, doi: 10.1016/j.ydbio.2004.03.031.

66. Davidson, E.H. 2006. *The Regulatory Genome: Gene Regulatory Networks in Development and Evolution.* Academic Press. Burlington, MA.

67. Levine, M. 2010. Transcriptional enhancers in animal development and evolution. *Curr. Biol.* **20:** R754–R763, doi: 10.1016/j.cub.2010.06.070.

68. Wittkopp, P.J. 2010. Variable transcription factor binding: a mechanism of evolutionary change. *PLoS Biol.* **8:** e1000342, doi: 10.1371/journal.pbio.1000342.

69. Janssen, R. & W.G. Damen. 2006. The ten Hox genes of the millipede Glomeris marginata. *Dev. Genes. Evol.* **216:** 451–465.

70. Hughes, C.L. & T.C. Kaufman. 2002. Exploring the myriapod body plan: expression patterns of the ten Hox genes in a centipede. *Development* **129:** 1225–1238.

71. Papillon, D. & M.J. Telford. 2007. Evolution of Hox3 and ftz in arthropods: insights from the crustacean *Daphnia pulex*. *Dev. Genes. Evol.* **217:** 315–322.

72. Nusslein-Volhard, C. & E. Wieschaus. 1980. Mutations affecting segment number and polarity in *Drosophila*. *Nature* **287:** 795–801.

73. Wakimoto, B.T. & T.C. Kaufman. 1981. Analysis of larval segmentation genotypes associated with the *Antennapedia* gene complex in *Drosophila melanogaster*. *Dev. Biol.* **81:** 51–64.

74. Wakimoto, B.T., F.R. Turner & T.C. Kaufman. 1984. Defects in embryogenesis in mutants associated with the antennapedia gene complex of *Drosophila melanogaster*. *Dev. Biol.* **102:** 147–172.

75. Hafen, E., A. Kuroiwa & W.J. Gehring. 1984. Spatial distribution of transcripts from the segmentation gene *fushi tarazu* during *Drosophila* embryonic development. *Cell* **37:** 833–841.

76. Kuroiwa, A., E. Hafen & W.J. Gehring. 1984. Cloning and transcriptional analysis of the segmentation gene *fushi tarazu* of *Drosophila*. **37:** 825–831.

77. Carroll, S.B. & M.P. Scott. 1985. Localization of the *fushi tarazu* protein during *Drosophila* embryogenesis. *Cell* **43:** 47–57.

78. Scott, M.P. & A.J. Weiner. 1984. Structural relationships among genes that control development: sequence homolgy between the *Antennapedia*, *Ultrabithorax*, and *fushi tarazu* loci of *Drosophila*. *Proc. Natl. Acad. Sci. USA* **81:** 4115–4119.

79. Struhl, G. 1985. Near-reciprocal phenotypes caused by inactivation or indiscriminate expression of the *Drosophila* segmentation gene *ftz*. **318:** 677–680.

80. Hughes, C.L., P.Z. Liu & T.C. Kaufman. 2004. Expression patterns of the rogue Hox genes *Hox3/zen* and *fushi tarazu* in the apterygote insect *Thermobia domestica*. *Evol. Dev.* **6:** 393–401.

81. Dawes, R., I. Dawson, F. Falciani, *et al.* 1994. *Dax*, a locust Hox gene related to fushi-tarazu but showing no pair-rule expression. *Development* **120:** 1561–1572.

82. Brown, S.J., R.B. Hilgenfeld & R.E. Denell. 1994. The beetle *Tribolium castaneum* has a fushi tarazu homolog expressed in stripes during segmentation. *Proc. Natl. Acad. Sci. USA* **91:** 12922–12926.

83. Dearden, P.K. *et al.* 2006. Patterns of conservation and change in honey bee developmental genes. *Genome Res.* **16:** 1376–1384.

84. Heffer, A., J. Shultz & L. Pick. 2010. Surprising flexibility in a conserved Hox transcription factor over 550 million years of evolution. *Proc. Natl. Acad. Sci. USA* **107:** 18040–18045.

85. Riley, P.D., S.B. Carroll & M.P. Scott. 1987. The expression and regulation of Sex combs reduced protein in *Drosophila* embryos. *Genes Dev.* **1:** 716–730.

86. Carroll, S.B., S.D. Weatherbee & J.A. Langeland. 1995. Homeotic genes and the regulation and evolution of insect wing number. *Nature* **375:** 58–61, doi: 10.1038/375058a0.

87. Rogers, B.T., M.D. Peterson & T.C. Kaufman. 1997. Evolution of the insect body plan as revealed by the Sex combs reduced expression pattern. *Development* **124:** 149–157.

88. Rogers, B.T. & T.C. Kaufman. 1996. Structure of the insect head as revealed by the EN protein pattern in developing embryos. *Development* **122:** 3419–3432.

89. DeCamillis, M.A., D.L. Lewis, S.J. Brown, *et al.* 2001. Interactions of the Tribolium Sex combs reduced and proboscipedia orthologs in embryonic labial development. *Genetics* **159:** 1643–1648.

90. Curtis, C.D. *et al.* 2001. Molecular characterization of Cephalothorax, the Tribolium ortholog of Sex combs reduced. *Genesis* **30:** 12–20.

91. Shippy, T.D., C.D. Rogers, R.W. Beeman, *et al.* 2006. The Tribolium castaneum ortholog of Sex combs reduced controls dorsal ridge development. *Genetics* **174:** 297–307, doi: 10.1534/genetics.106.058610.

92. Hughes, C.L. & T.C. Kaufman. 2000. RNAi analysis of Deformed, proboscipedia and Sex combs reduced in the milkweed bug Oncopeltus fasciatus: novel roles for Hox genes in the hemipteran head. *Development* **127:** 3683–3694.

93. Chesebro, J., S. Hrycaj, N. Mahfooz & A. Popadic. 2009. Diverging functions of Scr between embryonic and post-embryonic development in a hemimetabolous insect, Oncopeltus fasciatus. *Dev. Biol.* **329:** 142–151.

94. Passalacqua, K.D., S. Hrycaj, N. Mahfooz & A. Popadic. 2010. Evolving expression patterns of the homeotic gene Scr in insects. *Int. J. Dev. Biol.* **54:** 897–904, doi: 10.1387/ijdb.082839kp.

95. Popadic, A., A. Abzhanov, D. Rusch & T.C. Kaufman. 1998. Understanding the genetic basis of morphological evolution: the role of homeotic genes in the diversification of the arthropod bauplan. *Int. j. Dev. Biol.* **42:** 453–461.

96. Hrycaj, S., J. Chesebro & A. Popadic. 2010. Functional analysis of Scr during embryonic and post-embryonic development in the cockroach, Periplaneta americana. *Dev. Biol.* **341:** 324–334, doi: 10.1016/j.ydbio.2010.02.018.

97. Prud'homme, B. *et al.* 2011. Body plan innovation in treehoppers through the evolution of an extra wing-like appendage. *Nature* **473:** 83–86, doi: 10.1038/nature09977.

98. Wasik, B.R., D.J. Rose & A.P. Moczek. 2010. Beetle horns are regulated by the Hox gene, Sex combs reduced, in a species- and sex-specific manner. *Evol. Dev.* **12:** 353–362, doi: 10.1111/j.1525-142X.2010.00422.x.

99. Angelini, D.R., P.Z. Liu, C.L. Hughes & T.C. Kaufman. 2005. Hox gene function and interaction in the milkweed bug Oncopeltus fasciatus (Hemiptera). *Dev. Biol.* **287:** 440–455.

100. Saenko, S.V., M.S. Marialva & P. Beldade. 2011. Involvement of the conserved Hox gene Antennapedia in the development and evolution of a novel trait. *EvoDevo* **2:** 9, doi: 10.1186/2041-9139-2-9.

101. Weatherbee, S.D. *et al.* 1999. *Ultrabithorax* function in butterfly wings and the evolution of insect wing patterns. *Curr. Biol.* **9:** 109–115.

102. Brunetti, C.R. *et al.* 2001. The generation and diversification of butterfly eyespot color patterns. *Curr. Biol.* **11:** 1578–1585.

103. Greer, J.M., J. Puetz, K.R. Thomas & M.R. Capecchi. 2000. Maintenance of functional equivalence during paralogous *Hox* gene evolution. *Nature* **367:** 83–87.

104. Wu, S., G. Ying, Q. Wu & M.R. Capecchi. 2007. Toward simpler and faster genome-wide mutagenesis in mice. *Nature genet.* **39:** 922–930, doi: 10.1038/ng2060.

105. Tschopp, P., N. Fraudeau, F. Bena & D. Duboule. 2011. Reshuffling genomic landscapes to study the regulatory evolution of Hox gene clusters. *Pro. Natl. Acad. Sci. USA* **108:** 10632–10637, doi: 10.1073/pnas.1102985108.

106. Guo, T. *et al.* 2011. An evolving NGF-Hoxd1 signaling pathway mediates development of divergent neural circuits in vertebrates. *Nat. Neurosci.* **14:** 31–36, doi: 10.1038/nn.2710.

107. Krumlauf, R. 1994. *Hox* genes in vertebrate development. *Cell* **78:** 191–201.

108. McIntyre, D.C. *et al.* 2007. Hox patterning of the vertebrate rib cage. *Development* **134:** 2981–2989.

109. Cohn, M.J. & C. Tickle. 1999. Developmental basis of limblessness and axial patterning in snakes. *Nature* **399:** 474–479, doi: 10.1038/20944.

110. Di-Poi, N. *et al.* 2010. Changes in Hox genes' structure and function during the evolution of the squamate body plan. *Nature* **464:** 99–103, doi: 10.1038/nature08789.

111. Mansfield, J.H. & A. Abzhanov. 2010. Hox expression in the American alligator and evolution of archosaurian axial patterning. *J. Exp. Zool. B. Mol. Dev. Evol.* **314:** 629–644, doi: 10.1002/jez.b.21364.

112. Moczek, A.P., J. Andrews, T. Kijimoto, *et al.* 2007. Emerging model systems in evo-devo: horned beetles and the origins of diversity. *Evol. Dev.* **9:** 323–328, doi: 10.1111/j.1525-142X.2007.00168.x.

113. Moczek, A.P. 2008. On the origins of novelty in development and evolution. *BioEssays* **30:** 432–447, doi:10.1002/bies.20754.

114. Cifuentes, F.J. & A. Garcia-Bellido. 1997. Proximo-distal specification in the wing disc of Drosophila by the nubbin gene. *Pro. Natl. Acad. Sci. USA* **94:** 11405–11410.

115. Gompel, N., B. Prud'homme, P.J. Wittkopp, *et al.* 2005. Chance caught on the wing: cis-regulatory evolution and the origin of pigment patterns in *Drosophila*. *Nature* **433:** 481–487, doi: nature03235 [pii]10.1038/nature03235.

116. Prud'homme, B. *et al.* 2006. Repeated morphological evolution through cis-regulatory changes in a pleiotropic gene. *Nature* **440:** 1050–1053, doi: nature04597 [pii]10.1038/nature04597.

117. Wittkopp, P.J., B.K. Haerum & A.G. Clark. 2008. Independent effects of cis- and trans-regulatory variation on gene expression in *Drosophila melanogaster*. *Genetics* **178:** 1831–1835, doi: 10.1534/genetics.107.082032.

118. Mann, R. & S. Carroll. 2002. Molecular mechanisms of selector gene function and evolution. *Curr. Opin. Genet. Dev.* **12:** 592–600.

119. Hsia, C.C. & W. McGinnis. 2003. Evolution of transcription factor function. *Curr. Opin. Genet. Dev.* **13:** 199–206.

120. Wagner, G.P. & V.J. Lynch. 2008. The gene regulatory logic of transcription factor evolution. *Trends. Ecol. Evol.* **23:** 377–385.

121. Ronshaugen, M., N. McGinnis & W. McGinnis. 2002. Hox protein mutation and macroevolution of the insect body plan. *Nature* **415:** 914–917.

122. Galant, R. & S.B. Carroll. 2002. Evolution of a transcriptional repression domain in an insect Hox protein. *Nature* **415:** 910–913.

123. Vachon, G. *et al.* 1992. Homeotic genes of the Bithorax complex repress limb development in the abdomen of the *Drosophila* embryo through the target gene Distal-less. *Cell* **71:** 437–450.

124. Lohr, U., M. Yussa & L. Pick. 2001. *Drosophila fushi tarazu*: a gene on the border of homeotic function. *Curr. Biol.* **11:** 1403–1412.

125. Lohr, U. & L. Pick. 2005. Cofactor-interaction motifs and the cooption of a homeotic Hox protein into the segmentation pathway of *Drosophila melanogaster*. *Curr. Biol.* **15:** 643–649.

126. Heffer, A. & L. Pick. 2011. Rapid Isolation of Gene Homologs across Taxa: efficient identification and isolation of gene orthologs from non-model organism genomes, a technical report. *Evo Devo.* **2:** 7.

127. Woods, R.J. *et al.* 2011. Second-order selection for evolvability in a large *Escherichia coli* population. *Science* **331:** 1433–1436, doi: 10.1126/science.1198914.

128. Hanes, S.D. & R. Brent. 1989. DNA specificity of the bicoid activator protein is determined by homeodomain recognition helix residue 9. *Cell* **57:** 1275–1283.

129. Lynch, J. & C. Desplan. 2003. 'De-evolution' of Drosophila toward a more generic mode of axis patterning. *Int. J. Dev. Biol.* **47:** 497–503.

130. McGregor, A.P. 2005. How to get ahead: the origin, evolution and function of bicoid. *Bioessays* **27:** 904–913.

131. Lemke, S. *et al.* 2008. Bicoid occurrence and Bicoid-dependent hunchback regulation in lower cyclorrhaphan flies. *Evol. Dev.* **10:** 413–420, doi: 10.1111/j.1525-142X.2008.00252.x.

132. Lemke, S. *et al.* 2010. Maternal activation of gap genes in the hover fly Episyrphus. *Development* **137:** 1709–1719, doi: 10.1242/dev.046649.

133. Lynch, V.J. *et al.* 2004. Adaptive evolution of HoxA-11 and HoxA-13 at the origin of the uterus in mammals. *Proc. R. Soc. Lond. [Biol].* **271:** 2201–2207. doi: 10.1098/rspb.2004.2848

134. Lynch, V.J. *et al.* 2008. Adaptive changes in the transcription factor HoxA-11 are essential for the evolution of pregnancy in mammals. *Proc. Natl. Acad. Sci. USA* **105:** 14928–14933.

135. Lynch, V.J., K. Brayer, B. Gellersen & G.P. Wagner. 2009. HoxA-11 and FOXO1A cooperate to regulate decidual prolactin expression: towards inferring the core transcriptional regulators of decidual genes. *PLoS One* **4:** e6845, doi: 10.1371/journal.pone.0006845.

136. Brayer, K.J., V.J. Lynch & G.P. Wagner. 2011. Evolution of a derived protein-protein interaction between HoxA11 and Foxo1a in mammals caused by changes in intramolecular regulation. *Proc. Natl. Acad. Sci. USA* **108:** E414–E420, doi: 10.1073/pnas.1100990108.

137. Hsia, C.C., A.C. Pare, M. Hannon, *et al.* 2010. Silencing of an abdominal Hox gene during early development is correlated with limb development in a crustacean trunk. *Evol Dev* **12:** 131–143, doi: EDE399 [pii]10.1111/j.1525-142X.2010.00399.x.

138. Thomsen, S., G. Azzam, R. Kaschula, *et al.* 2010. Developmental RNA processing of 3′UTRs in Hox mRNAs as a context-dependent mechanism modulating visibility to microRNAs. *Development* **137:** 2951–2960, doi: 10.1242/dev.047324.

139. Patraquim, P., M. Warnefors & C.R. Alonso. 2011. Evolution of Hox post-transcriptional regulation by alternative polyadenylation and microRNA modulation within 12 Drosophila genomes. *Mol. Biol. Evol.* **28:** 2453–2460, doi: 10.1093/molbev/msr073.

140. Janssen, R. & G.E. Budd. 2010. Gene expression suggests conserved aspects of Hox gene regulation in arthropods and provides additional support for monophyletic Myriapoda. *EvoDevo* **1:** 4, doi: 10.1186/2041-9139-1-4.

141. Petruk, S. *et al.* 2006. Transcription of bxd noncoding RNAs promoted by trithorax represses Ubx in cis by transcriptional interference. *Cell* **127:** 1209–1221, doi: 10.1016/j.cell.2006.10.039.

142. Bender, W., B. Weiffenbach, F. Karch & M. Peifer. 1985. *Cold Spring Harbor Symposia on Quantitative Biology, Molecular Biology of Development.* **50:** 173–180. Cold Spring Harbor Press. Cold Spring Harbor, NY.

143. Hogness, D.S. *et al.* 1985. *Cold Spring Harbor Symposia on Quantitative Biology, Molecular Biology of Development.* **50:** 181–194. Cold Spring Harbor Press. Cold Spring Harbor, NY.

Ann. N.Y. Acad. Sci. ISSN 0077-8923

ANNALS OF THE NEW YORK ACADEMY OF SCIENCES
Issue: *The Year in Evolutionary Biology*

Inbreeding–stress interactions: evolutionary and conservation consequences

David H. Reed,[1]* Charles W. Fox,[2] Laramy S. Enders,[3] and Torsten N. Kristensen[4,5]

[1]Department of Biology, University of Louisville, Louisville, Kentucky. [2]Department of Entomology, University of Kentucky, Lexington, Kentucky. [3]Department of Entomology, University of Nebraska-Lincoln, Lincoln, Nebraska. [4]Department of Molecular Biology and Genetics, Aarhus University, Denmark. [5]Department of Bioscience, Aarhus University, Denmark

Address for correspondence: Charles W. Fox, Department of Entomology, University of Kentucky, Lexington, KY 40546–0091. cfox@uky.edu

The effect of environmental stress on the magnitude of inbreeding depression has a long history of intensive study. Inbreeding–stress interactions are of great importance to the viability of populations of conservation concern and have numerous evolutionary ramifications. However, such interactions are controversial. Several meta-analyses over the last decade, combined with omic studies, have provided considerable insight into the generality of inbreeding–stress interactions, its physiological basis, and have provided the foundation for future studies. In this review, we examine the genetic and physiological mechanisms proposed to explain why inbreeding–stress interactions occur. We specifically examine whether the increase in inbreeding depression with increasing stress could be due to a concomitant increase in phenotypic variation, using a larger data set than any previous study. Phenotypic variation does usually increase with stress, and this increase can explain some of the inbreeding–stress interaction, but it cannot explain all of it. Overall, research suggests that inbreeding–stress interactions can occur via multiple independent channels, though the relative contribution of each of the mechanisms is unknown. To better understand the causes and consequences of inbreeding–stress interactions in natural populations, future research should focus on elucidating the genetic architecture of such interactions and quantifying naturally occurring levels of stress in the wild.

Keywords: biodiversity conservation; environmental stress; evolution; omics; inbreeding

Introduction

Inbreeding and stressful environmental conditions are two major variables that influence the ecological and evolutionary dynamics of natural populations.[1–3] Inbreeding causes reduced fitness in inbred relative to outbred individuals (i.e., inbreeding depression) and exposure to abiotic and biotic stressors, by definition, also decreases fitness relative to benign environments.[4–7] Rapid changes to natural habitats that have been experienced by many plant and animal populations during the last century (e.g., due to climate change) often increase the level of stress perceived by individuals[8,9] and, at the same time, lead to a reduction in population size and increased rates of inbreeding. For the management of threatened wild and domesticated species, it is therefore crucial to understand how the combined effects of inbreeding and decreased environmental quality affect population fitness.[10] As a result, understanding the degree to which inbreeding depression changes with environmental conditions has become a central focus in evolution, ecology, conservation, and animal breeding research.

An important question that emerged in the literature is whether decreases in fitness are additive when inbreeding and stress are combined, or if fitness is decreased more (or less) than expected under the assumption that inbreeding and stress act independently. When the simultaneous effects of inbreeding and stressful environmental conditions are not additive, there is an inbreeding–stress interaction (Fig. 1). As we demonstrate is this review,

---

*David H. Reed passed away on 24 October 2011. David worked on this paper until his death. He will be missed as both a friend and terrific scientist.

doi: 10.1111/j.1749-6632.2012.06548.x

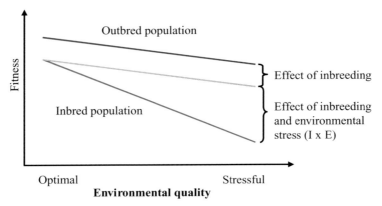

Figure 1. Fitness effects of inbreeding–environment interactions. Assuming the effect of inbreeding is independent of the environment, the reduction in fitness as a result of reduced environmental quality will be equal for outbred and inbred populations. The blue and gray lines illustrate fitness of an outbred and an inbred population, respectively, in the absence of inbreeding–environment interactions. Inbreeding depression is, however, often more severe under stressful environmental conditions. Thus, the red line illustrates fitness of an inbred population taking into account the effect of inbreeding–environment interactions (redrawn from Ref. 85).

inbreeding–stress interactions in which inbreeding depression increases under adverse environmental conditions are typically observed and have numerous repercussions for evolutionary biology and for the conservation of biodiversity. Nonadditive effects of these two sources of reduced population fitness, in which environmental stressors substantially increase the fitness consequences of inbreeding, can reduce thresholds for population persistence well below those predicted by models that assume that the effects of environmental stress and inbreeding are independent.[10,11]

Following the proxy that is widely used in studies of mutational effects, we here define stressfulness of an environment as a function of the mean fitness of outbred individuals in that environment relative to other environments.[4–7,12] Any environmental variable that reduces mean population fitness is thus considered a stressor. This includes ecological variables that increase physiological stress (e.g., as measured by increases in stress hormones or proteins), but only to the extent that increased physiological stress is associated with reduced fitness.

Historical development of inbreeding–stress research

Interest in inbreeding–stress interactions goes back at least 60 years. In the 1950s and 1960s, a number of papers, mostly in the agricultural literature, examined such interactions but generally lumped inbreeding–stress interactions together with the more general phenomena of genotype–environment interactions. Early papers[13–21] approached the problem from the perspective of Waddington and Lerner's ideas concerning heterozygosity, developmental stability, and maintenance of the optimum phenotype across changing environmental conditions (canalization).[22,23] Thus, the emphasis was on hybrid vigor upon crossing inbred lines of domesticated species, and not on the loss of genetic diversity and increased homozygosity due to habitat fragmentation and persistent small population size in natural environments. Surprisingly, despite increasing rates of inbreeding in many livestock breeds (which was intensified in the 1960s and 1970s with the development of reproductive technologies and advanced breeding schemes) and highly variable rearing conditions for these livestock, inbreeding–stress interactions were rarely investigated in the agricultural sciences during the last 50 years. Thus, there is a niche for developing models that incorporate inbreeding–environment interactions into quantitative genetic models used in breeding programs, analogous to that recently attempted for genotype–environment interactions,[24] with the potential to make breeding programs more effective.

While there was variation in the results among the early studies mentioned above, they were consistent enough for Wright[25] to suggest that less heterozygous (more inbred) individuals were generally more sensitive to environmental stress. However, it took a decade after Wright's suggestion before researchers again turned their attention in earnest to the question of the magnitude of inbreeding depression in stressful or variable environments. This renewed interest developed among evolutionary and conservation geneticists who sought to understand how inbreeding depression varied with environmental conditions, for example by comparing estimates from laboratory versus field or greenhouse conditions[26–30] and wild versus captive zoo populations.[31] Particularly influential in furthering ideas and producing copious data on inbreeding–stress interactions were Volker Loeschcke, Kuke Bijlsma and colleagues.[1,32–36] Now the question was refocused from one of heterozygote advantage to two important issues that persist in research pursuits today: (1) Are the effects of stress and inbreeding independent or are they synergistic? and (2) Are inbreeding effects general across different types of stressors or are they stress specific?

This renewed attention generated huge amounts of data on how inbreeding depression varies with environmental conditions but, due to substantial variation in results among studies, it did little to settle the basic question of whether inbreeding depression generally increases with stress. In 2005, a meta-analysis[5] confirmed Wright's intuition that environmental stress on average increased the magnitude of inbreeding depression. However, 24% of the studies showed no such increase with some even showing the opposite pattern (lower inbreeding depression in more stressful environments). Different species, populations, inbred lines, sexes, and families were highly variable in their response to inbreeding and stress. This variation helps explain why reaching any robust conclusion has been so difficult. At this point the general consensus was that while inbreeding depression often increased with stress, the specifics of the effects of stress on inbreeding depression were idiosyncratic to the genetic architecture of the population and the type of stress applied.

Despite the fact that the results from the meta-analysis by Armbruster and Reed[5] did not reveal evidence for a general mechanism underlying inbreeding–stress interactions, this paper spurred even more research investigating inbreeding–stress interactions, with more than 20 papers being published on the topic since 2005. In 2011, another meta-analysis was published, stimulated by a study of multiple levels of stress, mixing two different stressors (temperature and diet), on two populations of the seed-feeding beetle *Callosobruchus maculates*.[6] This meta-analysis found that much of the variation in the environmental impact on levels of inbreeding depression among studies could be explained by the amount of stress imposed; studies imposing very little stress tended to find no effect on inbreeding depression (Fig. 2A), whereas studies imposing severe stress found large effects of stress on inbreeding depression.[6] In this study, the inbreeding load, L, increased by about one lethal equivalent for each 30% difference in outbred fitness between environments. These results suggest that the effects of stress on inbreeding depression are more homogeneous than formerly thought, and not so idiosyncratic regarding the genetic architecture of the population or the type of environmental variable causing stressful conditions, with greater levels of stress consistently leading to more inbreeding depression. This result has been confirmed by an independent meta-analysis using *Drosophila* and experiments in both the laboratory and field.[37] Just as was found by Fox and Reed,[6] Enders and Nunney[37] found a strong linear relationship between inbreeding depression and the magnitude of multiple stressors, with inbreeding depression increasing linearly as the level of stress increases (Fig. 2B).

This review examines the hypotheses proposed to explain how inbreeding–stress interactions occur, current evidence to support these hypotheses, and what they mean for the evolution of small populations and conservation of biodiversity. We also make numerous suggestions concerning where future research in this field should be directed.

Why does inbreeding depression increase with stress?

A number of hypotheses have been proposed to explain the mechanisms by which stress can amplify levels of inbreeding depression.[2,38] In general, inbreeding–stress interactions can be viewed as resulting from (1) the effects of exposure to stress on the expression of deleterious alleles (focus is at the genetic level) and/or (2) the phenotypic effects caused by the expression of genetic load that affects resistance to stress (focus is at the phenotypic level). Here, we outline three major hypotheses proposed to explain inbreeding–stress interactions as well as current evidence to support each. It is important to note that these

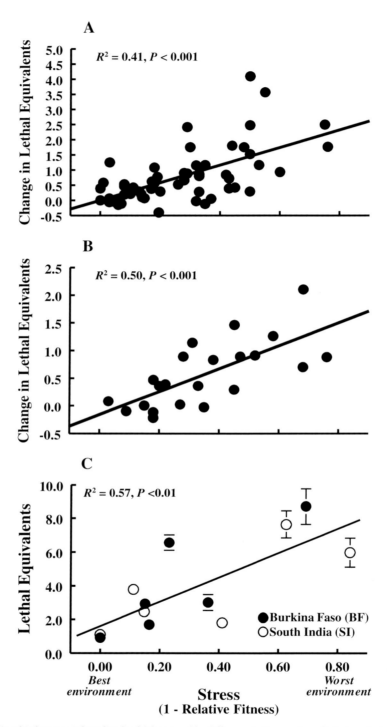

Figure 2. The relationship between inbreeding load (L) or L_{diff} (the difference in the number of lethal equivalents expressed in the stressful vs. benign environment) and the magnitude of stress in (A) a meta-analysis of published studies,[6] (B) a meta-analysis of *Drosophila* laboratory studies,[37] and (C) an experimental study of the beetle *Callosobruchus maculatus* reared at three temperatures on two host species.[6] Stress is calculated as $1 - \text{Survival}_{\text{outbred(stressful)}}/\text{Survival}_{\text{outbred(benign)}}$, and thus is by definition 0 in the most benign environment.

hypotheses are nonmutually exclusive and the relative contribution of each to inbreeding–stress interactions is currently unknown and should be the subject of future research. Moreover, because it is ultimately the expression of genetic load that will lead to the physiological and phenotypic changes that reduce the fitness of inbred individuals, the hypotheses presented are not entirely distinct (see discussion below), but have been organized according to whether the focus is the genetic (hypothesis 1) or phenotypic level (hypotheses 2 and 3).

Hypothesis 1: exposure to stressful environments alters the genetic architecture underlying inbreeding depression (i.e., the expression of genetic load)

The level of inbreeding depression is dependent, at least in part, on the expression of recessive deleterious alleles (genetic load)—specifically, the overall number of deleterious alleles expressed and the relative fitness effect of each expressed recessive allele.[2,35,38,39] Exposure to stressful environments can therefore lead to increased levels of inbreeding depression by affecting the expression of genetic load in two ways: (1) increasing fitness costs associated with particular deleterious alleles and/or (2) increasing the number of deleterious alleles expressed relative to those expressed in benign environments.

Inbreeding–stress interactions can occur when stress magnifies the average negative effect of deleterious recessive alleles, leading to increases in the selection coefficients against these alleles in stressful environments.[40–43] Environmental differences in natural selection (i.e., environment-dependent selection) are recognized as important sources of inbreeding–stress interactions, distinct from those mechanisms that contribute to environment-dependent phenotypic expression.[2] It is not necessary to assume genotype-by-environment interactions when changes in the intensity of selection against deleterious alleles contribute to increased levels of inbreeding depression, simply because the mortality of inbred individuals relative to outbred individuals increases as selection increases.[38]

Numerous studies demonstrate that different loci often affect the same trait in different environments (e.g., in QTL studies), and that the degree to which specific loci affect a trait varies with environmental conditions.[44,45] For such changes in effects of individual loci to generate an increase in inbreeding depression under stressful conditions requires that deleterious mutations—those that are generally recessive and thus exposed to selection by increased inbreeding—be, on average, disproportionately affected by stressful conditions. This is, however, inconsistent with the results of some models that predict that environmental conditions primarily change the variance of mutational fitness effects rather than their average effect or their net expression level. Stress should affect which alleles are expressed, and the variance in effect size among alleles, but not the average effect size of deleterious alleles, and thus not the average effect of the genetic load.[12]

Inbreeding–stress interactions may also result from specific genotype-by-environment interactions that arise through the expression of condition-dependent deleterious alleles that are neutral or beneficial under benign environments, but become deleterious under stress.[35,38] This explanation is distinct from the above hypothesis (where inbreeding–stress interactions magnify the average negative effect of deleterious recessive alleles) in that alleles that are neutral or beneficial in benign environments become detrimental under stressful environmental conditions; that is, there is a change in the sign of the selection coefficient on these recessive alleles. For example, Vermeulen and Bijlsma[46,47] demonstrated temperature-specific adult mortality in inbred *Drosophila melanogaster* lines caused by the expression of temperature-sensitive lethal alleles, alleles that were neutral or even beneficial at some temperatures but lethal at other temperatures. Condition-dependent deleterious alleles can be maintained in a population when purging is ineffective due to the infrequency with which organisms encounter stressful conditions, such as novel or particularly extreme conditions, that may be typically avoided due to habitat selection. Environmental-dependent deleterious alleles may contribute to the significant lineage effects observed under stress in many studies,[5,46,47] explaining in part why independent but equally inbred lines can behave very differently under stress. However, some evidence indicates that the effects of new mutations are highly and positively correlated across environmental conditions[48–52] (but see Refs. 53–55), and that inbreeding depression of genotypes is generally positively correlated across environmental conditions (e.g., Ref. 56 and references therein), suggesting that increased expression of condition-dependent deleterious alleles may not be the only mechanism causing an increase in inbreeding depression with stress. This is, however, in contrast to the finding of condition-dependent deleterious alleles being very important for explaining levels of inbreeding depression in lifespan and thermal tolerance in *D. melanogaster*.[46,47] Mutations segregating within populations appear to be far less positively correlated in their effects across environments,[5,57,58] as would be expected since selection against mutations that are deleterious in all environments should remove such mutations rapidly. Furthermore, Hillenmeyer *et al.*[59] found that 97% of genes that are required for optimal growth were environment-specific in yeast. In summary, there is support for the hypothesis that conditionally expressed recessive deleterious alleles partly explain inbreeding–environment interactions.

Hypothesis 2: the expression of genetic load increases the sensitivity of inbred individuals to the physiological effects of environmental stress

The expression of deleterious alleles is predicted to render inbred individuals more susceptible to the effects of environmental stress by causing overall physiological weakening and thus greater fitness costs relative to outbred individuals.[23,60,61] Increased sensitivity can result from changes in the expression of genetic load (as described in hypothesis 1) that adversely affect basic cellular functioning and ultimately influence an individual's phenotype and overall fitness. In general, disruption of the stress response system caused by the expression of genetic load is predicted to reduce or eliminate the ability of organisms to buffer their physiology and repair or reduce tissue and genomic damage experienced during exposure to stress. A growing body of literature suggests that deleterious mutations commonly decanalize the phenotype against random environmental perturbations, and thus increase the sensitivity of most traits to environmental perturbations.[52] The hypothesis that deleterious mutations decanalize the phenotype against environmental stress is indirectly supported by observations that inbreeding depression often increases with age[60,62] (but see Ref. 63). Assuming that inbreeding depression is caused by recessive deleterious alleles and

that natural selection acts more weakly against late-acting deleterious alleles (such that the expression of deleterious mutations increases with age, as predicted by mutation accumulation models of senescence), then inbreeding depression should increase with age. In this case, deleterious recessive alleles are expressed and thus decanalize the phenotype only at old age. Thus, inbreeding–age interactions share characteristics with inbreeding–environment interactions where the deleterious effect of certain recessive alleles is observed only under harsh environmental conditions. In vertebrates (and to a lesser degree other groups), inbreeding also directly increases expression of mutations that disrupt the generalized immune and stress response system. This has given rise to the hypothesis that inbreeding reduces an individual's ability to resist parasites and pathogens.[39]

Another proposed explanation for increased sensitivity of inbred individuals to stressful environments is that inbreeding hinders adaptive phenotypic plasticity.[38] Plasticity can be defined as the ability of a genotype to produce varied phenotypic outcomes depending on the environment. If plastic responses provide a short-term and partly "emergency" solution to cope with sudden changes in the environment, then a reduced ability of more homozygous individuals to exhibit plasticity in response to changes in the environment, especially if this occurs at the physiological level, may provide a general explanation for why environmental conditions that are harmless to outbred individuals could be perceived as highly stressful by inbred individuals. This hypothesis is supported by work in *Drosophila* demonstrating that inbreeding can reduce the capacity to maintain high fitness across environments[64,65] as well as recent data showing that inbreeding reduces the expression of predator-induced adaptive plasticity in shell thickness in a hermaphroditic snail species (*Physa acuta*).[66] However, studies across plants and animals examining the effect of inbreeding on plastic responses have shown varied results.[67] In general, a better understanding of the ability to react to environmental changes via adaptive phenotypic plasticity in small and fragmented natural populations exposed to ecologically relevant environmental variation is needed to verify the generality of the "inbreeding depression for plasticity" hypothesis. Future work is needed to determine how sensitive inbred populations in nature are to the effects of stress and to what extent inbreeding may impede the ability of such populations to cope with environmental change via plasticity and/or evolutionary adaptation.

Hypothesis 3: inbreeding depression under stress is the consequence of increased phenotypic variation

It has been proposed that inbreeding depression itself is a form of selection and therefore predicted to increase under stressful conditions that accentuate phenotypic variance.[61] The amount of phenotypic variance in a population sets a limit on the degree to which the fitness of distinct groups of individuals can differ. Crow[68] showed this for natural selection by demonstrating that the index of total selection (CV^2, the squared phenotypic coefficient of variation) sets a limit to how much selection can occur in a population (though Downhower *et al.*[69] caution that this index can be misleading when CV^2 and the mean are not independent of each other). Waller *et al.*[61] pointed out that CV^2 also constrains the magnitude of inbreeding depression that can occur in a population. If the amount of phenotypic variation present increases with the stressfulness of the environment, then the opportunity for fitness to differ between inbred and outbred individuals similarly increases, and so we might expect inbreeding depression to covary with the degree of stress. Thus, an increase in phenotypic variance with stress (Waller's hypothesis) is a mechanism by which stress can increase inbreeding depression; stress often increases phenotypic variation, and thus the slope of the relationship between stress and inbreeding depression, which is constrained by the relationship between stress and phenotypic variation, will increase with stress.

There are several mechanisms that could contribute to increased phenotypic variation under stressful conditions. As previously discussed, exposure to stress can decanalize growth and development, which has been shown to reveal cryptic genetic variation and give rise to the appearance of new phenotypes.[70,71] In *Drosophila* (flies) and *Danio* (zebra fishes), a reduced ability to buffer against the cellular effects of stress have been shown to cause increased morphological asymmetries and even lead to changes in the frequencies of novel phenotypes in laboratory populations.[72,73] Stressful conditions may therefore alter the expression of genetic load by

revealing underlying mutations that are otherwise hidden by normal physiological buffering, thus increasing the variance in fitness of both inbred and outbred individuals. Increased phenotypic variation could also result from the effects of stress on the regulation of gene expression, for example, by increasing transcriptional errors and introducing noise in expression.[74] There is some evidence suggesting that gene expression is more variable when individuals are exposed to stress and that stress-related genes exhibit high levels of noise relative to housekeeping genes,[75] increasing phenotypic variation in the population. However, it remains unclear if inbred individuals are more susceptible to the effects of stress on phenotypic variation and whether this may influence levels of inbreeding depression.

Evaluating the role of phenotypic variance in inbreeding–stress interactions

In general, it is unknown to what extent increased phenotypic variation under stress contributes to inbreeding–stress interactions. Waller and colleagues[61] found that CV^2 was a poor predictor of inbreeding depression for a given trait across abiotic and biotic stress treatments, but that levels of inbreeding depression were positively correlated with levels of phenotypic variability (CV^2) when considered across nine fitness-related traits measured in *Brassica rapa*. However, this study did not evaluate the role of stress level in the expression of inbreeding depression, which may explain why inbreeding depression was found to be positive, negative, or zero depending on the trait and stressor applied. Currently, there are no studies examining the role of both phenotypic variation and stress levels in determining the outcome of inbreeding–stress interactions. To test the relative importance of increasing phenotypic variance in generating observed inbreeding–stress interactions, we performed a multiple regression analysis on nine data sets (Tables 1 and S1). For each data set, the dependent variable is the number of lethal equivalents for each inbred line in each of the environments, which differed in stress levels. The independent variables are CV^2, degree of stress (decrement in relative fitness of the outbred population in each environment), and the interaction between the two.

We found that that stress increased CV^2 in eight of nine data sets, but the correlation between the degree of stress and the increase in CV^2 was weak

Table 1. The relative importance of stress, CV^2, and their interaction, in explaining variation in the inbreeding load across environments

Variable	Importance value (weight)
CV^2	0.630
Stress	0.862
CV^{2*} stress	0.226

NOTE: See Table S1 for details of the nine analyses that were included using model averaging to estimate importance values.

(mean correlation coefficient, $r = 0.34 \pm 0.12$) for all except one study.[6] The best-fit multiple regression model consistently explained significant amounts of the variation among inbred lines in their number of lethal equivalents ($P < 0.01$ in all cases, mean $R^2 = 0.53 \pm 0.03$). We then used an information-theoretic approach to select the best-fit model (Table S1) and used model averaging to weight the relative importance of CV^2, stress, and their interaction in determining inbreeding depression (Table 1).

The model including an effect of stress, but not CV^2, was the best-fit model in four of nine data sets, whereas the model including just CV^2 alone was the best fit in only one data set (Table S1). In the remaining data sets, both CV^2 and stress were important; for two data sets the best-fit model included stress and CV^2 and in two others the best-fit model included CV^2, stress, and the interaction between them. Stress effects independent of increases in CV^2 were the single most important variable, across models, determining the level of inbreeding depression (Table 1). However, CV^2 was similar in importance to stress level. Thus, the consensus strongly suggests that stress often increases inbreeding depression by increasing CV^2, but that the increase in CV^2 explains only part of the variance in inbreeding depression; there are also other independent mechanisms by which stress increases inbreeding depression. The interaction between CV^2 and the independent effects of stress is clearly not as important as the main effects (Table 1). However, the interaction term seems important in three of the nine data sets and is consistently negative.

These analyses were performed on a very limited subset of published data and on only a few study species. They are projects that included at least one of the authors of this paper as a coauthor and to

which we had unfettered access to the data. It is worth noting, however, that there are no consistent patterns among authors or organisms. For two species of spiders within the same genus, stress alone was the best-fit model for one species and the worst-fit model for the other species. One study, using a single population of *D. melanogaster*, found very different results depending on whether fecundity or egg-to-adult survival was used as a fitness surrogate. Thus, we expect these results to be fairly general.[76]

These findings differ from those of Waller *et al.*[61] where mixed support for CV^2 and no support for independent effects of stress were found. Differences in results may be due to the strength of the stress used in the studies and the amount of inbreeding depression the populations actually experience.

The physiological basis of inbreeding depression and inbreeding–environment interactions

Inbreeding itself can mimic environmental stress at the cellular level. Kristensen *et al.*[77] and Pedersen *et al.*[78] found increased expression levels of the stress-induced heat shock protein 70 in replicate inbred lines as compared with outbred lines of *D. melanogaster* and *D. buzzatii*. An increase in levels of heat shock proteins in inbred individuals may be a general phenomenon that is involved in buffering the effects of deleterious mutations on protein instability and misfolding; inbreeding increases expression of deleterious alleles that reduce protein stability and increase protein misfolding, which, in turn, induces upregulation of heat shock proteins.[79–81]

Consistent with the results showing upregulation of heat shock proteins in inbred lines, it has been found in full genome transcriptomics studies that inbreeding leaves a directional fingerprint on gene regulation across lineages of *D. melanogaster*.[82,83] Genes that respond transcriptionally to inbreeding are primarily involved in stress resistance, immunity, and fundamental metabolic processes. The transcriptomic analyses of inbred lines show that although the genetic causation of inbreeding depression is unique for every population, a general response can be identified that is likely to be explained by stress mechanisms being induced by inbreeding and not due to disruption of specific gene products (which would be lineage specific). This view is supported by metabolite profiling, which also reveals a clear separation of inbred and outbred lines.[84,85]

In summary, the available data from transcriptomic and metabolomic investigations of inbreeding effects demonstrate that inbreeding imposes physiological changes, as expected, given the clear reduction in fitness often observed in response to inbreeding. More unexpectedly, the data show that expression of the genetic load induces directional molecular responses, such as differential expression of major metabolic pathways and protein quality control systems that may counteract the deleterious effects of inbreeding. Most notable for our understanding of inbreeding–stress interactions is that many of the genes whose transcription responds to inbreeding are those involved in a variety of stress responses, including heat shock proteins and genes involved in immune processes, indicating that physiologically organisms respond to inbreeding as if they are being exposed to multisimultaneous environmental stressors.

Genome-wide transcriptome studies have also been used to describe how inbreeding–environment interactions manifest at the biochemical and physiological levels.[81] Kristensen *et al.*[81] showed that more genes were differentially expressed with inbreeding in *D. melanogaster* after exposure to temperature stress relative to benign conditions, signifying inbreeding–environment interactions. Transcripts involved in major metabolic pathways, in particular, were affected by the interaction. Thus, the sparse documentation of inbreeding–environment interactions on the transcript level suggests that at this molecular level inbreeding and the environment do not influence organisms additively.

Future perspectives using omics tools

The ability to investigate molecular phenotypes using omics technologies has been influential in expanding our knowledge about the effects of inbreeding and inbreeding–environment interactions. Nevertheless, the underlying molecular and biochemical mechanistic details of inbreeding effects are still unclear and there is a need for more hypothesis-driven investigations (e.g., using genetically modified organisms) in which the roles of specific genes, transcripts, proteins, and metabolites in inbred and outbred individuals are tested at different environmental conditions. Results of studies at the transcript level should be followed up by mechanistic studies that pinpoint the importance

of candidate genes and biochemical pathways for explaining inbreeding–environment interactions.

Genomic tools enabling the establishment of complete genome sequences, not only for model organisms but also for species of conservation interest, will enable researchers to perform genotyping at low cost for thousands of single-nucleotide polymorphism (SNP) markers.[86] Information on genome-wide SNPs can, for example, be useful in pinpointing the genetic basis of variation in inbreeding effects across environments, species, populations, and families. Potentially, information from genomic studies revealing recessive deleterious alleles of importance for inbreeding–environment interactions can be used to control recessive defects in captive populations by using this molecular information to select parents for the next generation.

Genomic information is currently being used intensively to guide selection decisions in animal and plant breeding, as it is expected that this will lead to faster rates of genetic improvement than does the use of traditional methods.[87] For example, methods are being developed that allow estimation of the level of inbreeding based on genomic information (personal communication, Louise Dybdahl Pedersen). This will enable a much more accurate estimate of inbreeding compared to estimates obtained based on pedigree information. Genome-wide SNP genotyping can therefore be used to precisely monitor and efficiently control the rate of inbreeding[88] in domesticated or managed wild populations. Genomic tools have the potential to allow control of inbreeding rates and heterozygosity at loci of crucial importance for fitness, which will allow fixation of favorable alleles in traits of importance for fitness while maintaining genetic variation in other parts of the genome. However, this field is in its infancy and the method described is obviously only of practical use in domesticated animals, zoo populations, plants in botanical garden, or otherwise heavily managed populations. Furthermore, for it to be efficient in relation to minimizing detrimental effects of inbreeding–environment interactions, genes/SNPs that govern inbreeding depression across environmental conditions should be identified.

Genomic approaches can potentially also be used to address basic questions about the molecular basis and genetic architecture of inbreeding depression.[89] For instance, is inbreeding depression caused by a few or many loci? And, how much of the inbreeding depression results from dominance, overdominance, or epistasis? Such knowledge is important for predicting the potential efficacy of purging, genomic selection, and assisted migration between populations.[38,86,90] If inbreeding depression is covered by a few loci of large effect,[91] and if inbreeding depression in benign and stressful environments is covered by some of the same genes, genomic selection might be effective in purging the genetic load.

Despite the fascinating prospect of employing genomic technologies in research to identify mechanisms responsible for inbreeding depression and the environmental dependency of inbreeding depression, it is also important to keep in mind challenges and limitations. First, the genetic architecture of inbreeding depression and inbreeding–environment interactions is likely complex and varies among species populations and individuals within populations.[38,92] Second, loci of importance for inbreeding depression will probably not be the same across environments.[5,57,58] Therefore, for genomic selection to be efficient in populations kept in zoos, botanical gardens, or in semicaptive environments, management practices should be developed that minimize adaptation to captivity and resample environmental conditions that the populations are likely to experience if translocated back to nature.[93] Third, threatened inbred populations will be small by definition. This will reduce power and thereby accuracy of the results and reduce the potential to select effectively against recessive deleterious alleles. Fourth, today only a few species have genomes that have been sequenced and reference genomes are available for an even smaller number of species. This means that for almost all species of conservation concern, we are still far from being able to do what we have suggested above. However, this is likely to change within the next 10 years with further developments in molecular biology, noninvasive sampling methods, and in bioinformatics.

The importance of inbreeding–stress interactions for conservation and evolutionary biology

We have defined a stressor as any environmental factor that reduces the fitness of an individual or population.[4–7,12] Populations in nature are constantly exposed to various forms of stress, such as pathogens and parasites, hunger and thirst, extreme heat or cold, toxic substances, and the risk of pre-

dation. Stress is likely particularly high in organisms of conservation interest because of anthropogenic activities that create novel or suboptimal conditions (e.g., global climate change, introduced species, pathogens, and pollution). For example, a growing body of literature demonstrates that individuals in fragmented or poor-quality habitats,[94,95] and those exposed to novel predators[96] or parasites,[97] express higher levels of stress hormones, indicating that they experience greater levels of physiological stress. Environmental stressors can induce physiological stress, such as changes in hormone levels, which can in turn lead to increased susceptibility to disease and predation, and/or generally reduce fitness.

Consequences of inbreeding–stress interactions for small populations

As environments continue to rapidly change worldwide, populations are not only subjected to progressively higher levels of stress in the form of industrial pollution, pesticides, and changes in ambient temperatures, but are also becoming increasingly smaller, more fragmented, and less genetically diverse. The increased risk of extinction due to the negative impacts of random genetic drift and inbreeding on disease resistance, evolutionary potential, and overall fitness are well established,[3,98–104] genetically depauperate populations have lower fitness, lowered disease resistance, and less evolutionary potential.[102,105] However, there is added risk for small populations when the deleterious effects of stress are amplified in inbred individuals. Simultaneous increases in stress and inbreeding rates and levels are thus expected to rapidly ratchet up extinction rates.[1,10,106] Extinction risk is going to be determined primarily by the extreme downturns in population size[102,107] and these will become more extreme than predicted by Liao and Reed[10] under the assumption that the interaction becomes stronger as stress becomes greater.[6,37]

Liao and Reed[10] determined that including reasonable estimates of the inbreeding–environment interaction reduces persistence times by 17.5–28.5% for a wide range of realistic assumptions about population dynamics and genetics and Robert[11] concluded that unbiased assessments of the viabilities of species is only obtained by identifying and integrating the most important processes governing persistence times (i.e., demography and genetics).

Liao and Reed[10] also identified some counterintuitive patterns; for example, the influence of the inbreeding–stress interaction on the median time to population extinction was greatest for larger populations. This is because populations currently viewed as relatively safe from extinction can more quickly cross the threshold into the extinction vortex when large inbreeding–stress interactions occur. Of course, this does not mean that inbreeding–environment interactions are insignificant for small populations. In contrast, although the proportional effect of the inbreeding–environment interaction may be less for small populations, such populations are already in crisis, and already experiencing the inbreeding conditions for which the interaction is important. The consequences of the interaction in increasing the risk of extinction are thus more imminent for smaller populations. Consideration of inbreeding–environment interactions in models of population persistence and conservation efforts should therefore be a priority.

Despite evidence from simulation studies and studies on organisms in the laboratory, it still remains to be shown in nature whether inbreeding–stress interactions do speed up extinction rates to the degree predicted based on studies that do not take into account all specific genetic details, such as the specifics of selection (purging, and balancing and directional selection). In addition, inbreeding is known to have multigenerational effects on fitness, and the same is certainly true for some types of stress,[108] but it is unclear whether the effects of inbreeding–stress interactions persist across generations. Low reproductive values persisting beyond the period of actual stress could prolong population recovery and increase the probability of entering an extinction vortex. Future studies are needed that examine the role of inbreeding–stress interactions under natural conditions, particularly in small and fragmented populations, with focus on the potential for multigenerational effects. Most laboratory studies can be criticized for not being ecologically relevant as they often investigate rather extreme levels of inbreeding and only one stressor (but see Ref. 6). This is problematic as the importance of inbreeding–stress interactions are depending on the level of inbreeding and expected to be more severe with exposure to multiple stresses that can interact in their effect on the phenotype.[109–112] We have only limited knowledge on such inbreeding–stress

interactions and future studies should also focus on natural populations or in laboratory studies investigating multiple environmental stresses.

Relevance of inbreeding–stress interactions for purging genetic load

Although environmental stress is commonly viewed as increasing inbreeding depression,[5,61] stress has also been proposed to increase selection against recessive deleterious alleles expressed in homozygous individuals, thus purging genetic load.[35,113] Exposure to stress over multiple generations is predicted to reduce inbreeding depression by decreasing the frequency of deleterious alleles in the population over multiple generations,[114–116] but can also have an effect within generations (intragenerational) if fitness correlations exist across multiple life history stages.[60] Purging of genetic load has been heavily studied,[98] yet we still have little idea whether the effects of purging are general versus environment specific or if different type of stress vary in their ability to purge genetic load.[35,117]

Inbreeding–stress interactions could, in theory, lead to very rapid purging of the deleterious alleles responsible for such interactions. However, specific stresses can increase, decrease, or have no effect on the magnitude of selection against mutations,[118] thus contributing to differences in the degree of purging across stress types. In addition, understanding the contribution of stress-specific versus stress-general genes or pathways to inbreeding–stress interactions is imperative to understanding the dynamics of purging in natural populations. The genomics work cited above for *Drosophila* suggests that many of the deleterious alleles affecting inbreeding depression do so through genes affecting generalized stress responses, but we have far too little data to generalize. The answer will have particularly significant consequences for our ability to extrapolate from results of laboratory studies to nature, and for predicting responses of populations bred and studied in captivity that are intended for reintroduction into natural, and generally more stressful, conditions. For example, we might predict that the consequences of inbreeding depression will be greatest in novel environmental conditions—those to which the organism is not adapted and in which they have not had an opportunity to purge their genetic load. Limiting adaptation to the captive environment, such as for *ex situ* populations intended for reintroduction, may warrant explicit attempts to limit inbreeding–stress interactions.

Future inbreeding–stress research

Much remains to be understood about the impacts of inbreeding–stress interactions for biodiversity conservation. There is still much unexplained variation in the magnitude of inbreeding depression expressed under stressful conditions, suggesting that additional factors may be important in explaining how stress and inbreeding interact in populations of conservation interest.[5,6,37,61] Identifying the types of stress that are more or less likely to induce such interactions will have direct application to species management. Identifying categories of stressors that do or do not trigger inbreeding–stress interactions may also help us to understand the genomic and proteomic underpinnings of such interactions. It is thus particularly important that more research be done on the effects of inbreeding and environmental stress in wild populations. Laboratory experiments can only go so far in mimicking the complex variety of stressors and stress levels faced by organisms and it is also important to impose realistic levels of inbreeding. How inbreeding–stress interactions affect population dynamics has rarely been studied in natural populations. Few studies have looked at temporal variation in levels of inbreeding depression in the wild[119–125] and only one has correlated seasonal changes in inbreeding depression with concurrent changes in levels of stress.[37] Studies on natural populations, in the field, are therefore crucial for extrapolating from the wide diversity of studies on model laboratory systems to natural systems of conservation importance.

Among the more important aspects of natural environments that we poorly understand is the frequency and magnitude of various stressors. Stress can come in the form of fluctuations in temperature, humidity, food availability, mating opportunities, and risk of predation. However, the extent to which stressors, such as these contribute to inbreeding–stress interactions is relatively unexplored in natural populations. In addition, it is unknown to what degree various stressors might be similar in plants, invertebrate animals, and vertebrate animals. If common stressors can be identified, it will allow us to examine whether negative genetic correlations generally exist between them. Negative genetic correlations to different stressors

can severely limit evolutionary potential and curtail population growth.[44,126–128] Genetic correlations for resistance to commonly encountered stressors with moderately strong selection should be mostly or entirely positive, as selection should strongly favor mutations with positive effects across several stressors. This will be particularly the case if a small set of generalized stress responses mediates fitness across a range of most commonly encountered stressors. However, many things might limit or prevent these positive correlations from evolving. Populations may be too small to generate and effectively fix such mutations, there may be physiological reasons for the negative genetic correlation, or there may be a negative temporal correlation between heritability for a trait and the strength of selection against that trait.[129] Under these conditions, inbreeding-stress interactions will likely lead to inefficient purging of the genetic load even in the environment the purging occurred in and lead to rapid fixation of potentially deleterious alleles for other forms of stress.

Acknowledgments

We thank the University of Kentucky Agricultural Research Station for funding C.W.F. We also thank the Danish Research Council and the Vice-Chancellor at Aarhus University for funding to T.N.K. (via a STENO stipend) and D.H.R. (during his stay at Aarhus University, where he wrote the first draft of this review). David Reed passed away on 24 October 2011. David worked on this paper until his death. He will be missed as both a friend and terrific scientific colleague.

Conflicts of interest

The authors declare no conflicts of interest.

Supporting information

Additional supporting information may be found in the online version of this article:

Table S1. Comparisons of AIC_c values for statistical models testing whether stress or CV^2 is a better predictor of the inbreeding load for multiple species studied by the authors. Models are ranked from the best (lowest AIC_c) to worst (highest AIC_c). The Δ_i is the difference between the best-fit model and the model being compared to the best. The w_i is the weight or likelihood of a model relative to other models.

Please note: Wiley-Blackwell is not responsible for the content or functionality of any supporting materials supplied by the authors. Any queries (other than missing material) should be directed to the corresponding author of the article.

References

1. Bijlsma, R. *et al.* 2000. Does inbreeding affect the extinction risk of small populations? Predictions from Drosophila. *J. Evol. Biol.* **13:** 502–514.
2. Cheptou, P.O. & K. Donohue. 2011. Environment-dependent inbreeding depression: its ecological and evolutionary significance. *New Phytol.* **189:** 395–407.
3. Frankham, R. 2005. Genetics and extinction. *Biol. Conserv.* **126:** 131–140.
4. Hoffmann, A.A. & P.A. Parsons. 1997. *Extreme Environmental Change and Evolution.* Cambridge. Cambridge University Press.
5. Armbruster, P. & D.H. Reed. 2005. Inbreeding depression in benign and stressful environments. *Heredity* **95:** 235–242.
6. Fox, C.W. & D.H. Reed. 2011. Inbreeding depression increases with environmental stress: an experimental study and meta-analysis. *Evolution* **65:** 246–258.
7. Bijlsma, R. & V. Loeschcke. 2005. Environmental stress, adaptation and evolution: an overview. *J. Evol. Biol.* **18:** 744–749.
8. Toohey, B.D. & G.A. Kendrick. 2007. Survival of juvenile Ecklonia radiata sporophytes after canopy loss. *J. Exp. Mar. Biol. Ecol.* **349:** 170–182.
9. Karsten, U. *et al.* 2001. Photosynthetic performance of *Arctic macroalgae* after transplantation from deep to shallow waters. *Oecologia* **127:** 11–20.
10. Liao, W. & D.H. Reed. 2009. Inbreeding-environment interactions increase extinction risk. *Anim. Conserv.* **12:** 54–61.
11. Robert, A. 2011. Find the weakest link. A comparison between demographic, genetic and demo-genetic metapopulation extinction times. *BMC Evol. Biol.* **11:** 260.
12. Martin, G. & T. Lenormand. 2006. The fitness effect of mutations across environments: a survey in light of fitness landscape models. *Evolution* **60:** 2413–2427.
13. Lewis, D. 1955. Gene interaction, environment and hybrid vigour. *Proc. Royal Soc. Lond. B* **144:** 178–185.
14. Maynard Smith, J. 1956. Acclimatization to high temperatures in inbred and outbred *Drosophila subobscura*. *J. Genet.* **54:** 497–505.
15. Parsons, P.A. 1959. Genotypic-environmental interactions for various temperatures in *Drosophila melanogaster*. *Genetics* **44:** 1325–1333.
16. Finlay, K.W. 1963. Adaptation: its measurement and significance in barley breeding. In *Proceedings of the 1st International Barley Genetics Symposium.* Wageningen, The Netherlands.
17. Griffing, B. & J. Langridge. 1963. Phenotypic stability of growth in the self-fertilised species *Arabidopsis thaliana*. *Stat. Genet. Plant Breed.: NAS-NRC* **982:** 368–394.
18. Hull, P. *et al.* 1963. A comparison of the interaction, with types of environment, of pure strains or strain crosses of poultry. *Genet. Res.* **4:** 370–381.
19. Bucio-Alanis, L. 1966. Environmental and genotype-environmental components of variability. *Heredity* **21:** 387–405.
20. Pederson, D.G. 1968. Environmental stress, heterozygote advantage and genotype-environment interaction in Arabidopsis. *Heredity* **23:** 127–138.
21. Jinks, J.L. & K. Mather. 1955. Stability in development of heterozygotes and homozygotes. *Proc. R. Soc. Lond. B* **143:** 561–578.
22. Waddington, C.H. 1942. Canalization of development and the inheritance of acquired characters. *Nature* **150:** 563–565.
23. Lerner, I.M. 1954. *Genetic Homeostasis.* Wiley. New York.
24. Windig, J.J. *et al.* 2011. Simultaneous estimation of genotype by environment interaction accounting for discrete and continuous environmental descriptors in Irish dairy cattle. *J. Dairy Sci.* **94:** 3137–3147.
25. Wright, S. 1977. *Evolution and the Genetics of Populations, Vol. 3. Experimental Results and Evolutionary Deductions.* University of Chicago Press. Chicago.
26. Dudash, M.R. 1990. Relative fitness of selfed and outcrossed progeny in a self-compatible, protandrous species, *Sabatia angularis* l (Gentianaceae): a comparison in three environments. *Evolution* **44:** 1129–1139.
27. Armbruster, P. *et al.* 2000. Equivalent inbreeding depression under laboratory and field conditions in a tree-hole-breeding mosquito. *Proc. R. Soc. Lond. Ser. B-Biol. Sci.* **267:** 1939–1945.
28. Jimenez, J.A. *et al.* 1994. An experimental study of inbreeding depression in a natural habitat. *Science* **266:** 271–273.
29. Chen, X.F. 1993. Comparison of inbreeding and outbreeding in hermaphroditic *Arianta arbustorum* (l) (land snail). *Heredity* **71:** 456–461.
30. Johnston, M.O. 1992. Effects of cross and self-fertilization on progeny fitness in *Lobelia cardinalis* and *L. siphilitica*. *Evolution* **46:** 688–702.
31. Crnokrak, P. & D.A. Roff. 1999. Inbreeding depression in the wild. *Heredity* **83:** 260–270.
32. Dahlgaard, J. *et al.* 1995. Heat-shock tolerance and inbreeding in Drosophila buzzatii. *Heredity* **74:** 157–163.
33. Dahlgaard, J. & V. Loeschcke. 1997. Effects of inbreeding in three life stages of *Drosophila buzzatii* after embryos were exposed to a high temperature stress. *Heredity* **78:** 410–416.
34. Hauser, T.P. & V. Loeschcke. 1996. Drought stress and inbreeding depression in *Lychnis flos-cuculi* (Caryophyllaceae). *Evolution* **50:** 1119–1126.
35. Bijlsma, R. *et al.* 1999. Environmental dependence of inbreeding depression and purging in *Drosophila melanogaster*. *J. Evol. Biol.* **12:** 1125–1137.
36. Kristensen, T.N. *et al.* 2003. Effects of inbreeding and environmental stress on fitness: using *Drosophila buzzatii* as a model organism. *Conserv. Genet.* **4:** 453–465.
37. Enders, L.S. & L. Nunney. 2012. Seasonal stress drives predictable changes in inbreeding depression in captive field populations of *Drosophila melanogaster*? In revision.
38. Bijlsma, R. & V. Loeschcke. 2012. Genetic erosion impedes adaptive responses to stressful environments. *Evol. App.* **5:** 117–129.

39. Keller, L.F. & D.M. Waller. 2002. Inbreeding effects in the wild. *Trends Ecol. Evol.* **17:** 230–241.

40. Lynch, M. & B. Walsh. 1998. *Genetics and Analysis of Quantitative Traits.* Sinauer. Sunderland, MA.

41. Yang, H.P. *et al.* 2001. Whole-genome effects of ethyl methanesulfonate-induced mutation on nine quantitative traits in outbred *Drosophila melanogaster. Genetics* **157:** 1257–1265.

42. Agrawal, A.F. & M.C. Whitlock. 2010. Environmental duress and epistasis: how does stress affect the strength of selection on new mutations. *Trends Ecol. Evol.* **25:** 450–458.

43. Laffafian, A. *et al.* 2010. Variation in the strength and softness of selection on deleterious mutations. *Evolution* **64:** 3232–3241.

44. Vieira, C. *et al.* 2000. Genotype-environment interaction for quantitative trait loci affecting life span in *Drosophila melanogaster. Genetics* **154:** 213–227.

45. Verhoeven, K.J.F. *et al.* 2008. Habitat-specific natural selection at a flowering-time QTL is a main driver of local adaptation in two wild barley populations. *Mol. Ecol.* **14:** 3416–3424.

46. Vermeulen, C.J. & R. Bijlsma. 2004a. Changes in mortality patterns and temperature dependence of lifespan in *Drosophila melanogaster* caused by inbreeding. *Heredity* **92:** 275–281.

47. Vermeulen, C.J. & R. Bijlsma. 2004b. Characterization of conditionally expressed mutants affecting age-specific survival in inbred lines of *Drosophila melanogaster*: lethal conditions and temperature-sensitive periods. *Genetics* **167:** 1241–1248.

48. Korona, R. 1999. Genetic load of the yeast *Saccharomyces cerevisiae* under diverse environmental conditions. *Evolution* **53:** 1966–1971.

49. Remold, S.K. & R.E. Lenski. 2001. Contribution of individual random mutations to genotype-by-environment interactions in *Escherichia coli. Proc. Natl. Acad. Sci. USA* **98:** 11388–11393.

50. Fry, J.D. & S.L. Heinsohn. 2002. Environment dependence of mutational parameters for viability in *Drosophila melanogaster. Genetics* **161:** 1155–1167.

51. Estes, S. *et al.* 2005. Spontaneous mutational correlations for life-history, morphological and behavioral characters in *Caenorhabditis elegans. Genetics* **170:** 645–653.

52. Baer, C.F. 2008. Quantifying the decanalizing effects of spontaneous mutations in Rhabditid nematodes. *Am. Nat.* **172:** 272–281.

53. Fry, J.D. *et al.* 1998. QTL mapping of genotype-environment interaction for fitness in *Drosophila melanogaster. Genet. Res.* **71:** 133–141.

54. Fernandéz, J. & C. López-Fanjul. 1996. Spontaneous mutational variances and covariances for fitness-related traits in *Drosophila melanogaster. Genetics* **143:** 829–837.

55. Fry, J.D. 2008. Genotype-environment interactions for fitness in Drosophila. *J. Genet.* **87:** 355–362.

56. Fox, C.W. *et al.* 2011. Inbreeding-environment interactions for fitness: complex relationships between inbreeding depression and temperature stress in a seed-feeding beetle. *Evol. Ecol.* **25:** 25–43.

57. Kondrashov, A.S. & D. Houle. 1994. Genotype-environment interactions and the estimation of the genomic mutation rate in *Drosophila melanogaster. Proc. R. Soc. Lond. B* **258:** 221–227.

58. Szafraniec, K. *et al.* 2001. Environmental stress and mutational load in diploid strains of the yeast Saccharomyces cerevisiae. *Proc. Natl. Acad. Sci. USA* **98:** 1107–1112.

59. Hillenmeyer, M.E. *et al.* 2008. The chemical genomic portrait of yeast: uncovering a phenotype for all genes. *Science* **320:** 362–365.

60. Enders, L.S. & L. Nunney. 2010. Sex-specific effects of inbreeding in wild-caught *Drosophila melanogaster* under benign and stressful conditions. *J. Evol. Biol.* **23:** 2309–2323.

61. Waller, D.M. *et al.* 2008. Effects of stress and phenotypic variation on inbreeding depression in *Brassica rapa. Evolution* **62:** 917–931.

62. Hughes, K.A. *et al.* 2002. A test of evolutionary theories of aging. *Proc. Natl. Acad. Sci. USA* **99:** 14286–14291.

63. Fox, C.W. *et al.* 2006. The genetic architecture of life span and mortality rates: gender and species differences in inbreeding load of two seed-feeding beetles. *Genetics* **174:** 763–773.

64. Fowler, K. & M.C. Whitlock. 1999. The distribution of phenotypic variance with inbreeding. *Evolution* **53:** 1143–1156.

65. Reed, D.H. *et al.* 2003. Fitness and adaptation in a novel environment: effect of inbreeding, prior environment, and lineage. *Evolution* **57:** 1822–1828.

66. Auld, J.R. & R.A. Relyea. 2010. Inbreeding depression in adaptive plasticity under predation risk in a freshwater snail. *Biol. Lett.* **6:** 222–224.

67. Kristensen, T.N. *et al.* 2011. No inbreeding depression for low temperature developmental acclimation across multiple Drosophila species. *Evolution* **65:** 3195–3201.

68. Crow, J.F. 1958. Some possibilities for measuring selection intensities in man. *Hum. Biol.* **30:** 1–13.

69. Downhower, J.F. *et al.* 1987. Opportunity for selection: an appropriate measure for evaluating variation in the potential for selection? *Evolution* **41:** 1395–1400.

70. Badyaev, A.V. & K.R. Foresman. 2004. Evolution of morphological integration. I. Functional units channel stress-induced variation in shrew mandibles. *Am. Nat.* **163:** 868–879.

71. Kassahn, K.S. *et al.* 2009. Animal performance and stress: responses and tolerance limits at different levels of biological organisation. *Biol. Rev.* **84:** 277–292.

72. Yeyati, P.L. *et al.* 2007. Hsp90 selectively modulates phenotype in vertebrate development. *PLoS Genet.* **3:** 431–447.

73. Rutherford, S.L. & S. Lindquist. 1998. Hsp90 as a capacitor for morphological evolution. *Nature* **396:** 336–342.

74. Raser, J.M. & E.K. O'Shea. 2005. Noise in gene expression: origins, consequences, and control. *Science* **309:** 2010–2013.

75. Lopez-Maury, L. *et al.* 2008. Tuning gene expression to changing environments: from rapid responses to evolutionary adaptation. *Nat. Rev. Genet.* **9:** 583–593.

76. Kristensen, T.N. *et al.* 2011. Slow inbred lines of *Drosophila melanogaster* express as much inbreeding depression as fast inbred lines under semi-natural conditions. *Genetica* **4:** 441–451.

77. Kristensen, T.N. *et al.* 2002. Inbreeding affects Hsp70 expression in two species of Drosophila even at benign temperatures. *Evol. Ecol. Res.* **4:** 1209–1216.

78. Pedersen, K.S. *et al.* 2005. Effects of inbreeding and rate of inbreeding in *Drosophila melanogaster*: Hsp70 expression and fitness. *J. Evol. Biol.* **18:** 756–762.

79. Sørensen, J.G. *et al.* 2003. The evolutionary and ecological role of heat shock proteins. *Ecol. Lett.* **6:** 1025–1037.

80. Paige, K.N. 2010. The functional genomics of inbreeding depression: a new approach to an old problem. *Bioscience* **60:** 267–277.

81. Kristensen, T.N. *et al.* 2006. Inbreeding by environmental interactions affect gene expression in *Drosophila melanogaster*. *Genetics* **173:** 1329–1336.

82. Kristensen, T.N. *et al.* 2005. Genome-wide analysis on inbreeding effects on gene expression in *Drosophila melanogaster*. *Genetics* **171:** 157–167.

83. Ayroles, J.F. *et al.* 2009. A genomewide assessment of inbreeding depression: gene number, function, and mode of action. *Conserv. Biol.* **23:** 920–930.

84. Pedersen, K.S. *et al.* 2010. Proteomic characterization of a temperature-sensitive conditional lethal in *Drosophila melanogaster*. *Heredity* **104:** 125–134.

85. Kristensen, T.N. *et al.* 2010. Research on inbreeding in the 'omic' era. *Trends Ecol. Evol.* **25:** 44–52.

86. Allendorf, F.W. *et al.* 2010. Genomics and the future of conservation genetics. *Nat. Rev. Genet.* **11:** 697–709.

87. Meuwissen, T. 2007. Genomic selection: marker assisted selection on a genome wide scale. *J. Anim. Breed. Genet.* **124:** 321–322.

88. Sonesson, A.K. *et al.* 2010. Maximising genetic gain whilst controlling rates of genomic inbreeding using genomic optimum contribution selection. In *Proceedings of the 9th World Congress on Genetics Applied to Livestock Production, 1–6 August 2010, No 892.* Leipzig, Germany.

89. Hagenblad, J. *et al.* 2009. Population genomics of the inbred Scandinavian wolf. *Mol. Ecol.* **18:** 1341–1351.

90. Sgrò, C.M. *et al.* 2011. Building evolutionary resilience for conserving biodiversity under climate change. *Evol. Appl.* **4:** 326–337.

91. Casellas, J. *et al.* 2009. Epistasis for founder-specific inbreeding depression in rabbits. *J. Hered.* **102:** 157–164.

92. Kristensen, T.N. *et al.* 2012. No inbreeding depression for low temperature developmental acclimation across multiple Drosophila species. In press.

93. Frankham, R. 2010. Challenges and opportunities of genetic approaches to biological conservation. *Biol. Conserv.* **143:** 1919–1927.

94. Suorsa, P. *et al.* 2003. Forest management is associated with physiological stress in an old–growth forest passerine. *Proc. R. Soc. Lond. B* **270:** 963–969.

95. Martínez-Mota, R. *et al.* 2007. Effects of forest fragmentation on the physiological stress response of black howler monkeys. *Anim. Conserv.* **10:** 374–379.

96. Berger, S. *et al.* 2007. Behavioral and physiological adjustments to new predators in an endemic island species, the Galápagos marine iguana. *Horm. Behav.* **52:** 653–663.

97. Martin, L.B. *et al.* 2010. The effects of anthropogenic global changes on immune functions and disease resistance. *Ann. NY Acad. Sci.* **1195:** 129–148.

98. Fox, C.W. *et al.* 2008. Experimental purging of the genetic load and its implications for the genetics of inbreeding depression. *Evolution* **62:** 2236–2249.

99. Willi, Y. & A.A. Hoffmann. 2009. Demographic factors and genetic variation influence population persistence under environmental change. *J. Evol. Biol.* **22:** 124–133.

100. Bouzat, J.L. 2010. Conservation genetics of population bottlenecks: the role of chance, selection, and history. *Conserv. Genet.* **11:** 463–478.

101. Ouborg, N.J. *et al.* 2010. Conservation genetics in transition to conservation genomics. *Trends Genet* **26:** 177–187.

102. Reed, D.H. 2010. Albatrosses, eagles, and newts, oh my!: exceptions to the prevailing paradigm concerning genetic diversity and population viability? *Anim. Conserv.* **13:** 448–457.

103. O'Grady, J.J. *et al.* 2006. Realistic levels of inbreeding depression strongly affect extinction risk in wild populations. *Biol. Conserv.* **133:** 42–51.

104. Blomqvist, D. *et al.* 2010. Trapped in the extinction vortex? Strong genetic effects in a declining vertebrate population. *BMC Evol. Biol.* **10:** 33.

105. Charpentier, M.J.E. *et al.* 2008. Inbreeding depression in ring-tailed lemurs (*Lemur catta*): genetic diversity predicts parasitism, immunocompetence, and survivorship. *Conserv. Genet.* **9:** 1605–1615.

106. Reed, D.H. *et al.* 2002. Inbreeding and extinction: the effect of environmental stress and lineage. *Conserv. Genet.* **3:** 301–307.

107. Reed, D.H. 2008. The effects of population size on population viability: from mutation to environmental catastrophes. In *Conservation Biology: Evolution in Action.* S.P. Carroll & C.W. Fox, Eds.: 16–34. Oxford University Press. New York.

108. Mousseau, T.A. & C.W. Fox. 1998. The adaptive significance of maternal effects. *Trends Ecol. Evol.* **13:** 403–407.

109. Halpern, B.S. *et al.* 2008. A global map of human impact on marine ecosystems. *Science* **319:** 948–952.

110. Crain, C.M. *et al.* 2008. Interactive and cumulative effects of multiple human stressors in marine systems. *Ecol. Lett.* **11:** 1304–1315.

111. Eranen, J.K. *et al.* 2009. Mountain birch under multiple stressors – heavy metal-resistant populations co-resistant to biotic stress but maladapted to abiotic stress. *J. Evol. Biol.* **22:** 840–851.

112. O'Donnell, M. *et al.* 2009. Predicted impact of ocean acidification on a marine invertebrate: elevated CO_2 alters response to thermal stress in sea urchin larvae. *Mar. Biol.* **156:** 439–446.

113. Leberg, P.L. & B.D. Firmin. 2008. Role of inbreeding depression and purging in captive breeding and restoration programmes. *Mol. Ecol.* **17:** 334–343.

114. Hedrick, P.W. 1994. Purging inbreeding depression and the probability of extinction: full-sib mating. *Heredity* **73:** 363–372.

115. Wang, J.L. *et al.* 1999. Dynamics of inbreeding depression due to deleterious mutations in small populations:

mutation parameters and inbreeding rate. *Genet. Res.* **74:** 165–178.

116. Wang, J.L. 2000. Effects of population structures and selection strategies on the purging of inbreeding depression due to deleterious mutations. *Genet. Res. Genet. Res.* **76:** 75–86.

117. Charlesworth, D. & J.H. Willis. 2009. The genetics of inbreeding depression. *Nat. Rev. Genet.* **10:** 783–796.

118. Agrawal, A.F. & M.C. Whitlock. 2010. Environmental duress and epistasis: how does stress affect the strength of selection on new mutations? *Trends Ecol. Evol.* **25:** 450–458.

119. Richardson, D.S. *et al.* 2004. Inbreeding in the Seychelles warbler: environment specific maternal effects. *Evolution* **58:** 2037–2048.

120. Rowe, G. & T.J.C. Beebee. 2005. Intraspecific competition disadvantages inbred natterjack toad (Bufo calamita) genotypes over outbred ones in a shared pond environment. *J. Anim. Ecol.* **74:** 71–76.

121. Marr, A.B. *et al.* 2006. Interactive effects of environmental stress and inbreeding on reproductive traits in a wild bird population. *J. Anim. Ecol.* **75:** 1406–1415.

122. Szulkin, M. & B. Sheldon. 2007. The environmental dependence of inbreeding depression in a wild bird population. *PLoS One* **2:** e1027.

123. Reed, D.H. *et al.* 2007. The genetic quality of individuals directly impacts population dynamics. *Anim. Conserv.* **10:** 275–283.

124. Keller, L.F. *et al.* 2002. Environmental conditions affect the magnitude of inbreeding depression in survival of Darwin's finches. *Evolution* **56:** 1229–1239.

125. Hayes, C.N. *et al.* 2005. Environmental variation influences the magnitude of inbreeding depression in Cucurbita pepo ssp texana (Cucurbitaceae). *J. Evol. Biol.* **18:** 147–155.

126. Weinig, C. *et al.* 2003. Heterogeneous selection at specific loci in natural environments in *Arabidopsis thaliana*. *Genetics* **165:** 321–329.

127. Zhong, D. *et al.* 2005. Costly resistance to parasitism: evidence from simultaneous quantitative trait mapping for resistance and fitness in *Tribolium castaneum*. *Genetics* **169:** 2127–2135.

128. Gardner, K.M. & R.G. Latta. 2007. Shared quantitative trait loci underlying the genetic correlation between continuous traits. *Mol. Ecol.* **16:** 4195–4209.

129. Wilson, A.J. *et al.* 2006. Environmental coupling of selection and heritability limits evolution. *PLoS Biology* **7:** 1270–1275.

Ann. N.Y. Acad. Sci. ISSN 0077-8923

ANNALS OF THE NEW YORK ACADEMY OF SCIENCES

Issue: *The Year in Evolutionary Biology*

The use of information theory in evolutionary biology

Christoph Adami[1,2,3]

[1]Department of Microbiology and Molecular Genetics, Michigan State University, East Lansing, Michigan. [2]Department of Physics and Astronomy, Michigan State University, East Lansing, Michigan. [3]BEACON Center for the Study of Evolution in Action, Michigan State University, East Lansing, Michigan

Address for correspondence: C. Adami, Department of Microbiology and Molecular Genetics, 2209 Biomedical and Physical Sciences Building, Michigan State University, East Lansing, MI 48824. adami@msu.edu

Information is a key concept in evolutionary biology. Information stored in a biological organism's genome is used to generate the organism and to maintain and control it. Information is also *that which evolves*. When a population adapts to a local environment, information about this environment is fixed in a representative genome. However, when an environment changes, information can be lost. At the same time, information is processed by animal brains to survive in complex environments, and the capacity for information processing also evolves. Here, I review applications of information theory to the evolution of proteins and to the evolution of information processing in simulated agents that adapt to perform a complex task.

Keywords: information theory; evolution; protein evolution; animat evolution

Introduction

Evolutionary biology has traditionally been a science that used observation and the analysis of specimens to draw inferences about common descent, adaptation, variation, and selection.[1,2] In contrast to this discipline that requires fieldwork and meticulous attention to detail, stands the mathematical theory of population genetics,[3,4] which developed in parallel but somewhat removed from evolutionary biology, as it could treat exactly only very abstract cases. The mathematical theory cast Darwin's insight about inheritance, variation, and selection into formulae that could predict particular aspects of the evolutionary process, such as the probability that an allele that conferred a particular advantage would go to fixation, how long this process would take, and how the process would be modified by different forms of inheritance. Missing from these two disciplines, however, was a framework that would allow us to understand the broad macro-evolutionary arcs that we can see everywhere in the biosphere and in the fossil record—the lines of descent that connect simple to complex forms of life. Granted, the existence of these unbroken lines—and the fact

that they are the result of the evolutionary mechanisms at work—is not in doubt. Yet, mathematical population genetics cannot quantify them because the theory only deals with existing variation. At the same time, the uniqueness of any particular line of descent appears to preclude a generative principle, or a framework that would allow us to understand the generation of these lines from a perspective once removed from the microscopic mechanisms that shape genes one mutation at the time. In the last 24 years or so, the situation has changed dramatically because of the advent of long-term evolution experiments with replicate lines of bacteria adapting for over 50,000 generations,[5,6] and *in silico* evolution experiments covering millions of generations.[7,8] Both experimental approaches, in their own way, have provided us with key insights into the evolution of complexity on macroscopic time scales.[6,8–14]

But there is a common concept that unifies the digital and the biochemical approach: information. That information is the essence of "that which evolves" has been implicit in many writings (although the word "information" does not appear in Darwin's *On the Origin of Species*). Indeed, shortly after the genesis of the theory of information at the hands of a Bell Laboratories engineer,[15] this theory was thought to ultimately explain everything from the higher functions of living organisms down to metabolism, growth, and differentiation.[16] However, this optimism soon gave way to a miasma of confounding mathematical and philosophical arguments that dampened enthusiasm for the concept of information in biology for decades. To some extent, evolutionary biology was not yet ready for a quantitative treatment of "that which evolves:" the year of publication of "Information in Biology"[16] coincided with the discovery of the structure of DNA, and the wealth of sequence data that catapulted evolutionary biology into the computer age was still half a century away.

Colloquially, information is often described as something that aids in decision making. Interestingly, this is very close to the mathematical meaning of "information," which is concerned with quantifying the ability to make predictions about uncertain systems. Life—among many other aspects— has the peculiar property of displaying behavior or characters that are appropriate, given the environment. We recognize this of course as the consequence of adaptation, but the outcome is that the adapted organism's decisions are "in tune" with its environment—the organism has *information* about its environment. One of the insights that has emerged from the theory of computation is that information must be physical—information cannot exist without a physical substrate that encodes it.[17] In computers, information is encoded in zeros and ones, which themselves are represented by different voltages on semiconductors. The information we retain in our brains also has a physical substrate, even though its physiological basis depends on the type of memory and is far from certain. Context-appropriate decisions require information, however it is stored. For cells, we now know that this information is stored in a cell's inherited genetic material, and is precisely the kind that Shannon described in his 1948 articles. If inherited genetic material represents information, then how did the information-carrying molecules acquire it? Is the amount of information stored in genes increasing throughout evolution, and if so, why? How much information does an organism store? How much in a single gene? If we can replace a discussion of the evolution of complexity along the various lines of descent with a discussion of the evolution of information, perhaps

then we can find those general principles that have eluded us so far.

In this review, I focus on two uses of information theory in evolutionary biology: First, the quantification of the information content of genes and proteins and how this information may have evolved along the branches of the tree of life. Second, the evolution of information-processing structures (such as brains) that control animals, and how the functional complexity of these brains (and how they evolve) could be quantified using information theory. The latter approach reinforces a concept that has appeared in neuroscience repeatedly: the value of information for an adapted organism is fitness,[18] and the complexity of an organism's brain must be reflected in how it manages to process, integrate, and make use of information for its own advantage.[19]

Entropy and information in molecular sequences

To define entropy and information, we first must define the concept of a *random variable*. In probability theory, a random variable X is a mathematical object that can take on a finite number of different *states* $x_1 \cdots x_N$ with specified probabilities $p_1, \ldots, p_N$. We should keep in mind that a mathematical random variable is a description—sometimes accurate, sometimes not—of a physical object. For example, the random variable that we would use to describe a fair coin has two states: $x_1 =$ heads and $x_2 =$ tails, with probabilities $p_1 = p_2 = 0.5$. Of course, an actual coin is a far more complex device—it may deviate from being true, it may land on an edge once in a while, and its faces can make different angles with true North. Yet, when coins are used for demonstrations in probability theory or statistics, they are most succinctly described with two states and two equal probabilities. Nucleic acids can be described probabilistically in a similar manner. We can define a nucleic acid random variable X as having four states $x_1 = A$, $x_2 = C$, $x_3 = G$, and $x_4 = T$, which it can take on with probabilities $p_1, \ldots, p_4$, while being perfectly aware that the nucleic acid molecules themselves are far more complex, and deserve a richer description than the four-state abstraction. But given the role that these molecules play as information carriers of the genetic material, this abstraction will serve us very well going forward.

Entropy

Using the concept of a random variable X, we can define its *entropy* (sometimes called *uncertainty*) as[20,21]

$$H(X) = -\sum_{i=1}^{N} p_i \log p_i. \qquad (1)$$

Here, the logarithm is taken to an arbitrary base that will normalize (and give units to) the entropy. If we choose the dual logarithm, the units are "bits," whereas if we choose base e, the units are "nats." Here, I will often choose the size of the alphabet as the base of the logarithm, and call the unit the "mer."[22] So, if we describe nucleic acid sequences (alphabet size 4), a single nucleotide can have up to 1 "mer" of entropy, whereas if we describe proteins (logarithms taken to the base 20), a single residue can have up to 1 mer of entropy. Naturally, a 5-mer has up to 5 mers of entropy, and so on.

A true coin, we can immediately convince ourselves, has an entropy of 1 bit. A single random nucleotide, by the same reasoning, has an entropy of 1 mer (or 2 bits) because

$$H(X) = -\sum_{i=1}^{4} 1/4 \log_4 1/4 = 1. \qquad (2)$$

What is the entropy of a nonrandom nucleotide? To determine this, we have to find the probabilities p_i with which that nucleotide is found at a particular position within a gene. Say we are interested in nucleotide 28 (counting from $5'$ to $3'$) of the 76 base pair tRNA gene of the bacterium *Escherichia coli*. What is its entropy? To determine this, we need to obtain an estimate of the probability that any of the four nucleotides are found at that particular position. This kind of information can be gained from sequence repositories. For example, the database tRNAdb[23] contains sequences for more than 12,000 tRNA genes. For the *E. coli* tRNA gene, among 33 verified sequences (for different anticodons), we find 5 that show an "A" at the 28th position, 17 have a "C," 5 have a "G," and 6 have a "T." We can use these numbers to estimate the substitution probabilities at this position as

$$p_{28}(A) = 5/33, \; p_{28}(C) = 17/33,$$
$$p_{28}(G) = 5/33, \; p_{28}(T) = 6/33, \qquad (3)$$

which, even though the statistics are not good, allow us to infer that "C" is preferred at that position.

The entropy of position variable X_{28} can now be estimated as

$$H(X_{28}) = -2 \times \frac{5}{33} \log_2 \frac{5}{33} - \frac{17}{33} \log_2 \frac{17}{33}$$
$$- \frac{6}{33} \log_2 \frac{6}{33} \approx 1.765 \text{ bits}, \qquad (4)$$

or less than the maximal 2 bits we would expect if all nucleotides appeared with equal probability. Such an uneven distribution of states immediately suggests a "betting" strategy that would allow us to make predictions with accuracy better than chance about the state of position variable X_{28}: If we bet that we would see a "C" there, then we would be right over half the time on average, as opposed to a quarter of the time for a variable that is evenly distributed across the four states. In other words, information is stored in this variable.

Information

To learn how to quantify the amount of information stored, let us go through the same exercise for a different position (say, position 41[a]) of that molecule, to find approximately

$$p_{41}(A) = 0.24, \; p_{41}(C) = 0.46,$$
$$p_{41}(G) = 0.21, \; p_{41}(T) = 0.09, \qquad (5)$$

so that $H(X_{41}) \approx 1.765$ bits. To determine how likely it is to find any particular nucleotide at position 41 *given* position 28 is a "C," for example, we have to collect *conditional* probabilities. They are easily obtained if we know the joint probability to observe any of the 16 combinations AA...TT at the two positions. The conditional probability to observe state j at position 41 given state i at position 28 is

$$p_{i|j} = \frac{p_{ij}}{p_j}, \qquad (6)$$

where p_{ij} is the *joint* probability to observe state i at position 28 and at the same time state j at position 41. The notation "$i \mid j$" is read as "i given j." Collecting these probabilities from the sequence data gives the probability matrix that relates the random

---

[a]The precise numbering of nucleotide positions differs between databases.

variable X_{28} to the variable X_{41}:

$p(X_{41}|X_{28})$

$$= \begin{pmatrix} p(A|A) & p(A|C) & p(A|G) & p(A|T) \\ p(C|A) & p(C|C) & p(C|G) & p(C|T) \\ p(G|A) & p(G|C) & p(G|G) & p(G|T) \\ p(T|A) & p(T|C) & p(T|G) & p(T|T) \end{pmatrix}$$

$$= \begin{pmatrix} 0.2 & 0.235 & 0 & 0.5 \\ 0 & 0.706 & 0.2 & 0.333 \\ 0.8 & 0 & 0.4 & 0.167 \\ 0 & 0.059 & 0.4 & 0 \end{pmatrix}. \qquad (7)$$

We can glean important information from these probabilities. It is clear, for example, that positions 28 and 41 are not independent from each other. If nucleotide 28 is an "A," then position 41 can only be an "A" or a "G," but mostly (4/5 times) you expect a "G." But consider the dependence between nucleotides 42 and 28

$$p(X_{42}|X_{28}) = \begin{pmatrix} 0 & 0 & 0 & 1 \\ 0 & 0 & 1 & 0 \\ 0 & 1 & 0 & 0 \\ 1 & 0 & 0 & 0 \end{pmatrix}. \qquad (8)$$

This dependence is striking—if you know position 28, you can predict (based on the sequence data given) position 42 with certainty. The reason for this perfect correlation lies in the functional interaction between the sites: 28 and 42 are paired in a stem of the tRNA molecule in a Watson–Crick pair—to enable the pairing, a "G" must be associated with a "C," and a "T" (encoding a U) must be associated with an "A." It does not matter which is at any position as long as the paired nucleotide is complementary. And it is also clear that these associations are maintained by the selective pressures of Darwinian evolution—a substitution that breaks the pattern leads to a molecule that does not fold into the correct shape to efficiently translate messenger RNA into proteins. As a consequence, the organism bearing such a mutation will be eliminated from the gene pool. This simple example shows clearly the relationship between information theory and evolutionary biology: Fitness is reflected in information, and when selective pressures maximize fitness, information must be maximized concurrently.

We can now proceed and calculate the information content. Each column in Eq. (7) represents a conditional probability to find a particular nucleotide at position 41, given a particular value is found at position 28. We can use these values to calculate the conditional entropy to find a particular nucleotide, given that position 28 is "A," for example, as

$$H(X_{41}|X_{28} = A)$$
$$= -0.2\log_2 0.2 - 0.8\log_2 0.8 \approx 0.72 \text{ bits}. \quad (9)$$

This allows us to calculate the amount of information that is revealed (about X_{41}) by knowing the state of X_{28}. If we do not know the state of X_{28}, our uncertainty about X_{41} is 1.795 bits, as calculated earlier. But revealing that X_{28} actually is an "A" has reduced our uncertainty to 0.72 bits, as we saw in Eq. (9). The information we obtained is then just the difference

$$I(X_{41} : X_{28} = A) = H(X_{41}) - H(X_{41}|X_{28} = A)$$
$$\approx 1.075 \text{ bits}, \qquad (10)$$

that is, just over 1 bit. The notation in Eq. (10), indicating information between two variables by a colon (sometimes a semicolon) is conventional. We can also calculate the *average* amount of information about X_{41} that is gained by revealing the state of X_{28} as

$$I(X_{41} : X_{28}) = H(X_{41}) - H(X_{41}|X_{28})$$
$$\approx 0.64 \text{ bits}. \qquad (11)$$

Here, $H(X_{41}|X_{28})$ is the average conditional entropy of X_{41} given X_{28}, obtained by averaging the four conditional entropies (for the four possible states of X_{28}) using the probabilities with which X_{28} occurs in any of its four states, given by Eq. (3). If we apply the same calculation to the pair of positions X_{42} and X_{28}, we should find that knowing X_{28} reduces our uncertainty about X_{42} to zero—indeed, X_{28} carries perfect information about X_{42}. The covariance between residues in an RNA secondary structure captured by the mutual entropy can be used to predict secondary structure from sequence alignments alone.[24]

Information content of proteins

We have seen that different positions within a biomolecule can carry information about other positions, but how much information do they store about the *environment* within which they evolve? This question can be answered using the same information-theoretic formalism introduced earlier. Information is defined as a reduction in our uncertainty (caused by our ability to make predictions with an accuracy better than chance) when armed with information. Here we will use proteins as our biomolecules, which means our random variables can take on 20 states, and our protein variable will be given by the joint variable

$$X = X_1 X_2 \cdots X_L, \quad (12)$$

where L is the number of residues in the protein. We now ask: "How much information *about the environment* (rather than about another residue) is stored in a particular residue?" To answer this, we have to first calculate the uncertainty about any particular residue in the absence of information about the environment. Clearly, it is the environment within which a protein finds itself that constrains the particular amino acids that a position variable can take on. If I do not specify this environment, there is nothing that constrains any particular residue i, and as a consequence the entropy is maximal

$$H(X_i) = H_{\max} = \log_2 20 \approx 4.32 \text{ bits.} \quad (13)$$

In any functional protein, the residue is highly constrained, however. Let us imagine that the possible states of the environment can be described by a random variable E (that takes on specific environmental states e_j with given probabilities). Then the information about environment $E = e_j$ contained in position variable X_i of protein X is given by

$$I(X_i : E = e_j) = H_{\max} - H(X_i \mid E = e_j), \quad (14)$$

in perfect analogy to Eq. (10). How do we calculate the information content of the entire protein, armed only with the information content of residues? If residues do not interact (that is, the state of a residue at one position does not reveal any information about the state of a residue at another position), then the information content of the protein would just be a sum of the information content of each residue

$$I(X : E = e_j) = \sum_{i=1}^{L} I(X_i : E = e_j). \quad (15)$$

This independence of positions certainly could not be assumed in RNA molecules that rely on

Watson–Crick binding to establish their secondary structure. In proteins, correlations between residues are much weaker (but certainly still important, see, e.g., Refs. 25–33), and we can take Eq. (15) as a first-order approximation of the information content, while keeping in mind that residue–residue correlations encode important information about the stability of the protein and its functional affinity to other molecules. Note, however, that a population with two or more subdivisions, where each subpopulation has different amino acid frequencies, can mimic residue correlations on the level of the whole population when there are none on the level of the subpopulations.[34]

For most cases that we will have to deal with, a protein is only functional in a very defined cellular environment, and as a consequence the conditional entropy of a residue is fixed by the substitution probabilities that we can observe. Let us take as an example the rodent homeodomain protein,[35] defined by 57 residues. The environment for this protein is of course the rodent, and we might surmise that the information content of the homeodomain protein in rodents is different from the homeodomain protein in primates, for example, simply because primates and rodents have diverged about 100 million years ago,[36] and have since then taken independent evolutionary paths. We can test this hypothesis by calculating the information content of rodent proteins and compare it to the primate version, using substitution probabilities inferred from sequence data that can be found, for example, in the Pfam database.[37] Let us first look at the entropy *per residue*, along the chain length of the 57 mer. But instead of calculating the entropy in bits (by taking the base-2 logarithm), we will calculate the entropy in "mers," by taking the logarithm to base 20. This way, a single residue can have at most 1 mer of entropy, and the 57-mer has at most 57 mers of entropy. The entropic profile (entropy per site as a function of site) of the rodent homeodomain protein depicted in Figure 1 shows that the entropy varies considerably from site to site, with strongly conserved and highly variable residues.

When estimating entropies from finite ensembles (small number of sequences), care must be taken to correct for the bias that is inherent in estimating the probabilities from the frequencies. Rare residues will be assigned zero probabilities in small ensembles but not in larger ones. Because this error is not

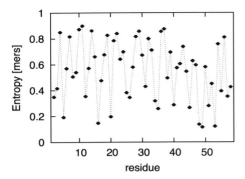

Figure 1. Entropic profile of the 57-amino acid rodent homeodomain, obtained from 810 sequences in Pfam (accessed February 3, 2011). Error of the mean is smaller than the data points shown. Residues are numbered 2–58 as is common for this domain.[35]

symmetric (probabilities will always be underestimated), the bias is always toward smaller entropies. Several methods can be applied to correct for this, and I have used here the second-order bias correction, described for example in Ref. 38. Summing up the entropies per site shown in Figure 1, we can get an estimate for the information content by applying Eq. (15). The maximal entropy $H_{\max}$, when measured in mers, is of course 57, so the information content is just

$$I_{\text{Rodentia}} = 57 - \sum_{i=1}^{57} H(X_i \mid e_{\text{Rodentia}}), \quad (16)$$

which comes out to

$$I_{\text{Rodentia}} = 25.29 \pm 0.09 \, \text{mers}, \quad (17)$$

where the error of the mean is obtained from the theoretical estimate of the variance given the frequency estimate.[38]

The same analysis can be repeated for the primate homeodomain protein. In Figure 2, we can see the difference between the "entropic profile" of rodents and primates

$$\Delta \text{Entropy} = H(X_i \mid e_{\text{Rodentia}}) - H(X_i \mid e_{\text{Primates}}), \quad (18)$$

which shows some significant differences, in particular, it seems, at the edges between structural motifs in the protein.

When summing up the entropies to arrive at the total information content of the primate home-

odomain protein we obtain

$$I_{\text{Primates}} = 25.43 \pm 0.08 \, \text{mers}, \quad (19)$$

which is identical to the information content of rodent homeodomains within statistical error. We can thus conclude that although the information is encoded somewhat differently between the rodent and the primate version of this protein, the total information content is the same.

Evolution of information

Although the total information content of the homeodomain protein has not changed between rodents and primates, what about longer time intervals? If we take a protein that is ubiquitous among different forms of life (i.e., its homologue is present in many different branches), has its information content changed as it is used in more and more complex forms of life? One line of argument tells us that if the function of the protein is the same throughout evolutionary history, then its information content should be the same in each variant. We saw a hint of that when comparing the information content of the homeodomain protein between rodents and primates. But we can also argue instead that because information is measured relative to the environment the protein (and thus the organism) finds itself in, then organisms that live in very different environments can potentially have different information content even if the sequences encoding the proteins are homologous. Thus, we could expect differences in protein information content in organisms that are different enough that the protein is used in different ways. But it is certainly not clear whether we should observe a trend of increasing or decreasing information along the line of descent. To get a first glimpse at what these differences could be like, I will take a look here at the evolution of information in two proteins that are important in the function of most animals—the homeodomain protein and the COX2 (cytochrome-c-oxidase subunit 2) protein.

The homeodomain (or homeobox) protein is essential in determining the pattern of development in animals—it is crucial in directing the arrangement of cells according to a particular body plan.[39] In other words, the homeobox determines where the head goes and where the tail goes. Although it is often said that these proteins are specific to animals, some plants have homeodomain proteins that are

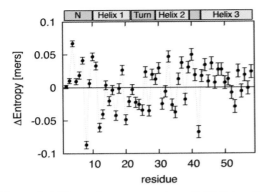

Figure 2. Difference between entropic profile "ΔEntropy" of the homeobox protein of rodents and primates (the latter from 903 sequences in Pfam, accessed February 3, 2011). Error bars are the error of the mean of the difference, using the average of the number of sequences. The colored boxes indicate structural domains as determined for the fly version of this gene. ("N" refers to the protein's "N-terminus").

homologous to those I study here.[40] The COX2 protein, on the other hand, is a subunit of a large protein complex with 13 subunits.[41] Whereas a nonfunctioning (or severely impaired) homeobox protein certainly leads to aborted development, an impaired COX complex has a much less drastic effect—it leads to mitochondrial myopathy due to a cytochrome oxidase deficiency,[42] but is usually not fatal.[43] Thus, by testing the changes within these two proteins, we are examining proteins with very different selective pressures acting on them.

To calculate the information content of each of these proteins along the evolutionary line of descent, in principle we need access to the sequences of extinct forms of life. Even though the resurrection of such extinct sequences is possible in principle[44] using an approach dubbed "paleogenetics,"[45,46] we can take a shortcut by grouping sequences according to the depth that they occupy within the phylogenetic tree. So when we measure the information content of the homeobox protein on the taxonomic level of the family, we include in there the sequences of homeobox proteins of chimpanzees, gorillas, and orangutans along with humans. As the chimpanzee version, for example, is essentially identical with the human version, we do not expect to see any change in information content when moving from the species level to the genus level. But we can expect that by grouping the sequences on the family level (rather than the genus or species

level), we move closer toward evolutionarily more ancient proteins, in particular because this group (the great apes) is used to reconstruct the sequence of the ancestor of that group. The great apes are but one family of the order *primates* which besides the great apes also contains the families of monkeys, lemurs, lorises, tarsiers, and galagos. Looking at the homeobox protein of all the primates then takes us further back in time. A simplified version of the phylogeny of animals is shown in Figure 3, which shows the hierarchical organization of the tree.

The database Pfam uses a range of different taxonomic levels (anywhere from 12 to 22, depending on the branch) defined by the NCBI Taxonomy Project,[47] which we can take as a convenient proxy for taxonomic depth—ranging from the most basal taxonomic identifications (such as phylum) to the most specific ones. In Figure 4, we can see the total sequence entropy

$$H_k(X) = \sum_{i=1}^{57} H(X_i|e_k), \qquad (20)$$

for sequences with the NCBI taxonomic level k, as a function of the level depth. Note that sequences at level k always include all the sequences at level $k–1$. Thus, $H_1(X)$, which is the entropy of all homeodomain sequences at level $k = 1$, includes the sequences of all eukaryotes. Of course, the taxonomic level description is not a perfect proxy for time. On the vertebrate line, for example, the genus *Homo* occupies level $k = 14$, whereas the genus *Mus* occupies level $k = 16$. If we now plot $H_k(X)$ versus k (for the major phylogenetic groups only), we see a curious splitting of the lines based only on total sequence entropy, and thus information (as information is just $I = 57 - H$ if we measure entropy in mers). At the base of the tree, the metazoan sequences split into chordate proteins with a lower information content (higher entropy) and arthropod sequences with higher information content, possibly reflecting the different uses of the homeobox in these two groups. The chordate group itself splits into mammalian proteins and the fish homeodomain. There is even a notable split in information content into two major groups within the fishes.

The same analysis applied to subunit II of the COX protein (counting only 120 residue sites that have sufficient statistics in the database) gives a very

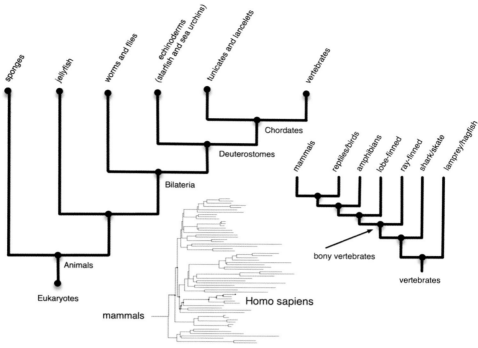

Figure 3. Simplified phylogenetic classification of animals. At the root of this tree (on the left tree) are the eukaryotes, but only the animal branch is shown here. If we follow the line of descent of humans, we move on the branch toward the vertebrates. The vertebrate clade itself is shown in the tree on the right, and the line of descent through this tree follows the branches that end in the mammals. The mammal tree, finally, is shown at the bottom, with the line ending in *Homo sapiens* indicated in red.

different picture. Except for an obvious split of the bacterial version of the protein and the eukaryotic one, the total entropy markedly decreases across the lines as the taxonomic depth increases. Furthermore, the arthropod COX2 is more entropic than the vertebrate one (see Fig. 5) as opposed to the ordering for the homeobox protein. This finding suggests that the evolution of the protein information content is specific to each protein, and most likely reflects the adaptive value of the protein for each family.

Evolution of information in robots and animats

The evolution of information within the genes of adapting organisms is but one use of information theory in evolutionary biology. Just as anticipated in the heydays of the "Cybernetics" movement,[48] information theory has indeed something to say about the evolution of information processing in animal brains. The general idea behind the connection between information and function is simple: Because information (about a particular system) is

what allows the bearer to make predictions (about that particular system) with accuracy better than chance, information is valuable as long as prediction is valuable. In an uncertain world, making accurate predictions is tantamount to survival. In other words, we expect that information, acquired from the environment and processed, has survival value and therefore is selected for in evolution.

Predictive information

The connection between information and fitness can be made much more precise. A key relation between information and its value for agents that survive in an uncertain world as a consequence of their actions in it was provided by Ay *et al.*,[49] who applied a measure called "predictive information" (defined earlier by Bialek *et al.*[50] in the context of dynamical systems theory) to characterize the behavioral complexity of an autonomous robot. These authors showed that the mutual entropy between a changing world (as represented by changing states in an organism's sensors) and the actions of motors that drive the agent's behavior (thus changing the future perceived states) is equivalent to Bialek's

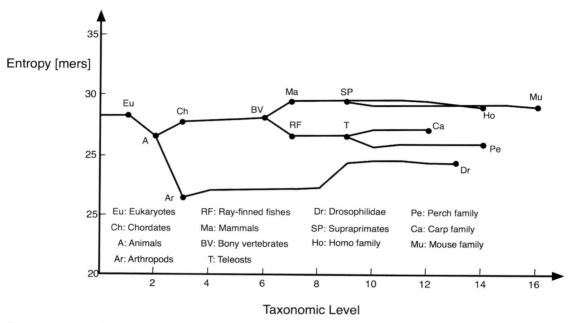

Figure 4. Entropy of homeobox domain protein sequences (PF00046 in the Pfam database, accessed July 20, 2006) as a function of taxonomic depth for different major groups that have at last 200 sequences in the database, connected by phylogenetic relationships. Selected groups are annotated by name. Fifty-seven core residues were used to calculate the molecular entropy. Core residues have at least 70% sequence in the database.

predictive information as long as the agent's decisions are Markovian, that is, only depend on the state of the agent and the environment at the preceding time. This predictive information is defined as the shared entropy between motor variables Y_t and the sensor variables at the subsequent time point X_{t+1}

$$I_{\text{pred}} = I(Y_t : X_{t+1}) = H(X_{t+1}) - H(X_{t+1}|Y_t). \quad (21)$$

Here, $H(X_{t+1})$ is the entropy of the sensor states at time $t + 1$, defined as

$$H(X_{t+1}) = -\sum_{x_{t+1}} p(x_{t+1}) \log p(x_{t+1}), \quad (22)$$

using the probability distribution $p(x_{t+1})$ over the sensor states x_{t+1} at time $t + 1$. The conditional entropy $H(X_{t+1}|Y_t)$ characterizes how much is left uncertain about the future sensor states X_{t+1} given the robot's actions in the present, that is, the state of the motors at time t, and can be calculated in the standard manner[20,21] from the joint probability distribution of present motor states and future sensor states $p(x_{t+1}, y_t)$.

As Eq. (21) implies, the predictive information measures how much of the entropy of sensorial

states—that is, the uncertainty about what the detectors will record next—is explained by the motor states at the preceding time point. For example, if the motor states at time t perfectly predict what will appear in the sensors at time $t + 1$, then the predictive information is maximal. Another version of the predictive information studies not the effect the motors have on future sensor states, but the effect the sensors have on future motor states instead, for example to guide an autonomous robot through a maze.[51] In the former case, the predictive information quantifies how actions change the perceived world, whereas in the latter case the predictive information characterizes how the perceived world changes the robot's actions. Both formulations, however, are equivalent when taking into account how world and robot states are being updated.[51] Although it is clear that measures such as predictive information should increase as an agent or robot learns to behave appropriately in a complex world, it is not at all clear whether information could be used as an objective function that, if maximized, will lead to appropriate behavior of the robot. This is the basic hypothesis of Linsker's "Infomax" principle,[52] which posits that neural control structures

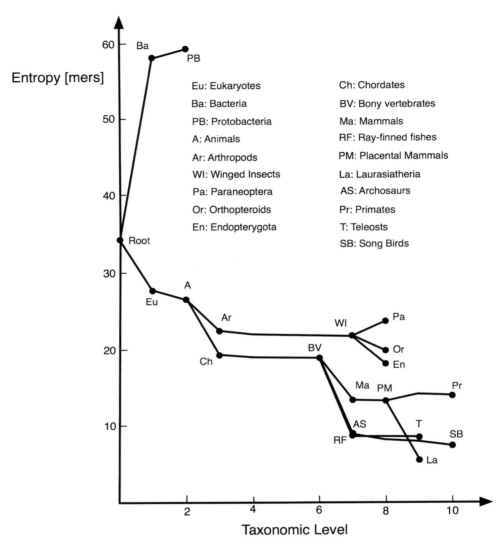

Figure 5. Entropy of COX subunit II (PF00116 in the Pfam database, accessed June 22, 2006) protein sequences as a function of taxonomic depth for selected different groups (at least 200 sequences per group), connected by phylogenetic relationships. One hundred twenty core residues were used to calculate the molecular entropy.

evolve to maximize "information preservation" subject to constraints. This hypothesis implies that the infomax principle could play the role of a guiding force in the organization of perceptual systems. This is precisely what has been observed in experiments with autonomous robots evolved to perform a variety of tasks. For example, in one task visual and tactile information had to be integrated to grab an object,[53] whereas in another, groups of five robots were evolved to move in a coordinated fashion[54] or else to navigate according to a map.[55] Such ex-

periments suggest that there may be a deeper connection between information and fitness that goes beyond the regularities induced by a perception–action loop, that connects fitness (in the evolutionary sense as the growth rate of a population) directly to information.

As a matter of fact, Rivoire and Leibler[18] recently studied abstract models of the population dynamics of evolving "finite-state agents" that optimize their response to a changing environment and found just such a relationship. In such a description, agents respond to a changing environment with a probability distribution $\pi(\sigma_t \mid \sigma_{t-1})$ of changing from state σ_{t-1} to state σ_t, to maximize the growth rate of the population. Under fairly general assumptions, the growth rate is maximized if the Shannon information that the agents can extract from the changing environment is maximal.[18] For our purposes, this Shannon information is nothing but the predictive information discussed earlier (see supplementary text S1 in Ref. 51 for a discussion of that point). However, such a simple relationship only holds if each agent perceives the environment in the same manner, and if information is acquired *only* from the environment. If information is inherited or retrieved from memory, on the other hand, predictive information cannot maximize fitness. This is easily seen if we consider an agent that makes decisions based on a combination of sensory input and memory. If memory states (instead of sensor states) best predict an agent's actions, the correlation between sensors and motors may be lost even though the actions are appropriate. A typical case would be navigation under conditions when the sensors do not provide accurate information about the environment, but the agent has nevertheless learned the required actions "by heart." In such a scenario, the predictive information would be low because the actions do not correlate with the sensors. Yet, the fitness is high because the actions were controlled by memory, not by the sensors. Rivoire and Leibler show further that if the actions of an agent are always optimal, given the environment, then a different measure maximizes fitness, namely the shared entropy between sensors and variables *given* the previous time step's sensor states[b]

$$I_{\text{causal}} = I(X_t : Y_{t+1} \mid X_{t-1}). \qquad (23)$$

In most realistic situations, however, optimal navigation strategies cannot be assumed. Indeed, as optimal strategies are (in a sense) the goal of evolutionary adaptation, such a measure could conceivably only apply at the endpoint of evolution. Thus, no general expression can be derived that ties these informational quantities directly to fitness.

---

[b]The notation is slightly modified here to conform to the formalism used in Ref. 51.

Integrated information

What are the aspects of information processing that distinguish complex brains from simple ones? Clearly, processing large amounts of information is important, but a large capacity is not necessarily a sign of high complexity. It has been argued that a hallmark of complex brain function is its ability to integrate disparate streams of information and mold them into a complex *gestalt* that represents more than the sum of its parts.[56–65] These streams of information come not only from different sensorial modalities such as vision, sound, and olfaction, but also (and importantly) from memory, and create a conscious experience in our brains that allows us to function at levels not available to purely reactive brains. One way to measure how much information a network processes is to calculate the shared entropy between the nodes at time t and time $t + 1$

$$I_{\text{total}} = I(Z_t : Z_{t+1}). \qquad (24)$$

Here, Z_t represents the state of the entire network (not just the sensors or motors) at time t, and thus the total information captures information processing among all nodes of the network, and can in principle be larger or smaller than the predictive information that only considers processing between sensor and motors.

We can write the network random variable Z_t as a product of the random variables that describe each node i, that is, each neuron, as (n is the number of nodes in the network)

$$Z_t = Z_t^{(1)} Z_t^{(2)} \cdots Z_t^{(n)}, \qquad (25)$$

which allows us to calculate the amount of information processed by each individual node i as

$$I^{(i)} = I\left(Z_t^{(i)} : Z_{t+1}^{(i)}\right). \qquad (26)$$

Note that I omitted the index t on the left-hand side of Eqs. (24) and (26), assuming that the dynamics of the network becomes stationary as $t \to \infty$, and, thus, that a sampling of the network states at any subsequent time points becomes representative of the agent's behavior. If the nodes in the network process information independently from each other, then the sum (over all neurons) of the information processed by each neuron would equal the amount of information processed by the entire network. The difference between the two then represents the amount of information that the network processes over and above the information processed by the individual neurons, the *synergistic informa-tion*[51]

$$SI_{\text{atom}} = I(Z_t : Z_{t+1}) - \sum_{i=1}^{n} I^{(i)} \left(Z_t^{(i)} : Z_{t+1}^{(i)} \right).$$
(27)

The index "atom" on the synergistic information reminds us that the sum is over the indivisible ele-ments of the network—the neurons themselves. As we see later, other more general partitions of the network are possible, and often times more appro-priate to capture synergy. The synergistic informa-tion is related to other measures of synergy that have been introduced independently. One is simply called "integration" and defined in terms of Shan-non entropies as[64,66,67]

$$\mathcal{I} = \sum_{i=1}^{n} H \left(Z_t^{(i)} \right) - H(Z_t).$$
(28)

This measure has been introduced earlier under the name "multi-information."[68,69] Another mea-sure, called Φ_{atom} in Ref. 51, was independently in-troduced by Ay and Wennekers[70,71] as a measure of the complexity of dynamical systems they called "stochastic interaction," and is defined as

$$\Phi_{\text{atom}} = \sum_{i=1}^{n} H \left(Z_t^{(i)} | Z_{t+1}^{(i)} \right) - H \left(Z_t | Z_{t+1} \right).$$
(29)

Note the similarity between Eqs. (27)–(29): whereas (27) measures synergistic information, (28) measures "synergistic entropy" and (29) synergistic conditional entropy in turn. The three are related because entropy and information are related, as for example in Eqs. (11) and (21). Using this relation, it is easy to show that[51]

$$\Phi_{\text{atom}} = SI_{\text{atom}} + \mathcal{I}.$$
(30)

Although we can expect that measures such as Eqs. (28)–(30) quantify some aspects of information in-tegration, it is likely that they overestimate the in-tegration because it is possible that elements of the computation are performed by groups of neurons that together behave as a single entity. In that case, subdividing the whole network into independent neurons may lead to the double counting of inte-grated information. A cleaner (albeit computation-ally much more expensive) approach is to find a partition of the network into nonoverlapping groups of nodes (parts) that are as independent of each other (information theoretically speaking) as possible. If we define the partition P of a network into k parts via $P = \{P^{(1)}, P^{(2)}, \ldots, P^{(k)}\}$, where each $P^{(i)}$ is a part of the network (a nonempty set of neurons with no overlap between the parts), then we can define a quantity that is analogous to Eq. (29) except that the sum is over the parts rather than the individual neurons[61]

$$\Phi(P) = \sum_{i=1}^{n} H \left(P_t^{(i)} | P_{t+1}^{(i)} \right) - H(P_t | P_{t+1}).$$
(31)

In Eq. (31), each part carries a time label because every part takes on different states as time proceeds. The so-called "minimum information partition" (or MIP) is found by minimizing a *normalized* Eq. (31) over all partitions

$$\text{MIP} = \arg \min_{P} \frac{\Phi(P_t)}{N(P_t)},$$
(32)

where the normalization $N(P_t) = (k - 1) \min_i [H_{\max}(P_t^{(i)})]$ balances the parts of the partition.[62] Using this MIP, the integrated informa-tion Φ is then simply given by

$$\Phi = \Phi (P = \text{MIP}).$$
(33)

Finally, we need to introduce one more concept to measure information integration in realistic evolv-ing networks. Because Φ of a network with a single (or more) disconnected nodes vanishes (because the MIP for such a network is always the partition into the connected nodes in one part, and the discon-nected nodes in another), we should attempt to define the computational "main complex," which is that sub-set of nodes for which Φ is maximal.[62] This measure will be called Φ_{MC} hereafter.

Although all these measures attempt to capture synergy, it is not clear whether any of them corre-late with fitness when an agent evolves, that is, it is not clear whether synergy or integration capture an aspect of the functional complexity of control structures that goes beyond the predictive infor-mation defined earlier. To test this, Edlund *et al.* evolved animats that learned, over 50,000 genera-tions of evolution, to navigate a two-dimensional maze,[51] constructed in such a way that optimal nav-igation requires memory. While measuring fitness, they also recorded six different candidate measures coevolutionary history, may vary greatly in magnitude in space and time, and may be embedded in a community of other interactors. Recognizing this rich detail is critical to understanding how and when mutualisms arise, persist, and break down. However, it is simulaneously important not to become lost in system-specific details. The framework of consumer–resource interactions provides a powerful unifying approach to tackle general issues across the diversity of mutualisms.

Competition is central to our understanding of consumer–resource interactions. Treating mutualisms as consumer–resource interactions (e.g., Refs. 2 and 4) thus places competition at the core of processes that shape the ecology and evolution of mutualists. The best documented interplay between competition and mutualism is that there is competition for the commodities mutualists produce; gaining access to mutualists in turn can change competitive outcomes. There are a number of distinct ways in which competition for mutualistic resources can occur and can influence the ecological and evolutionary dynamics of mutualism. For example, there may or may not be a predictable hierarchy of competitive dominance among potential partners,[5] or there may be a range of situations in which competitors actually benefit each other (via shared attraction of partners) rather than interfere.[6] Competition thus lies at the heart of selection to attract, retain, and benefit from mutualisms. However, although it is widely recognized that organisms compete for mutualists, the strength and consistency of such competition are rarely measured (but see Refs. 7–10). Indeed, it is often more an assumption than fact that competition for mutualists exists and can drive the evolution of traits that attract and reward partners.

Competition interacts with mutualism in other ways as well. It can occur between mutualistic and nonmutualistic (exploitative) partners for a shared resource; mutualism can involve putative competitors; and competitive advantage can arise as a benefit conferred by mutualists. The diverse ways in which competition and mutualism interact has been given surprisingly little focused attention. On the theoretical side, competition is generally present within models of ecological and evolutionary dynamics of mutualism; however, with relatively few exceptions (e.g., Refs. 11–14), its role is not much remarked upon. Palmer *et al.* provided an outstanding overview of the ways in which competition mediates mutualist coexistence,[15] but their perspective was largely descriptive and ecological. Here, we will argue that the existence of competition and of a competitive hierarchy among partners is critical to the outcome of mutualism at both ecological and evolutionary scales.

The structure of the paper is as follows. First, we synthesize the empirical literature on how competitive interactions are embedded within mutualism. To do this, we develop a novel graphical approach, built on the consumer–resource approach to mutualism. We then build on these ecological models to explore evolutionary aspects of the mutualism/competition interplay. We focus on the consequences and evolution of partner control mechanisms, then the factors fostering the evolution, persistence, and diversification of mutualism. We show how advancing our understanding of these issues requires knowledge of how competitive forces underlie mutualism. Finally, we identify intriguing evolutionary questions lying at the intersection of mutualism and competition that our approach could in the future be used to explore.

Competition and mutualism: empirical phenomena

Competition for mutualistic resources

Since mutualisms are defined as interactions that confer reciprocal benefits to two species, they are widely depicted as shown in Figure 1A. The partners are linked by positively labeled arrows, the top arrow in the figure indicating that species M1 confers benefits to M2, and the bottom arrow indicating that M2 confers benefits to M1. This abstraction of mutualism hides as much as it reveals, however. What are these benefits and is the exchange as simple and reciprocal as the figure might imply?

Figure 1A illustrates the *effects* of a single bout of commodity exchange, but not its underlying *mechanism*. While not shown here, competition can be included quantitatively in net effects diagrams through path coefficients.[16] Stanton's use of path analysis to quantify the interactions between and within guilds of mutualists was an important step in extending the study of mutualism beyond pairs of species. However, path analysis is still a top-down, phenomenological approach that does not identify the causes of the measured effects. In contrast, in Figure 1B, we try to capture more of the

A

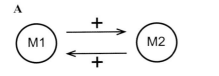

B

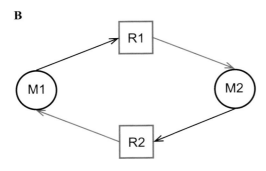

C

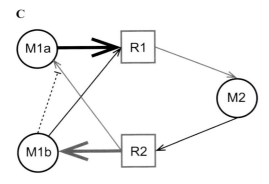

mechanism of commodity exchange underlying mutualism through a bottom-up depiction. Now, instead of the arrows connecting two species, the arrows point to a commodity that is then delivered to the partner. These commodities may be, in the language used above, either rewards or services; we group these into a single category, called *resources* (R). While we recognize their differences, we take this approach because in the context of exchange, they share two critical features: rewards and services both can be costly to offer to partners, and both can be competed for. In Figure 1B, there are still two species (M1 and M2), but now there are also two resources, R1 and R2; the arrows now are resource flows rather than (as in Fig. 1A) net effects. Black arrows indicate resources produced by a mutualist, and red arrows those that are consumed (either actually or metaphorically) by a mutualist. Thus, in Figure 1B, M1 produces R1, a resource that is consumed by M2; M2 in turn produces R2, a resource that is consumed by M1. As an example, M1 might be a plant that produces nectar (R1), which is consumed by M2, a bee, that produces the resource of pollen transport (R2) that is used profitably by M1.

A major advantage to depicting mutualism as shown in Figure 1B rather than Figure 1A is that it makes clear how competition can lie at the heart of these interactions. In Figure 1A, it is not evident what could be competed for; in Figure 1B, it is clear that it is the resources, R1 and R2. Indeed, the best documented way in which there is interplay between competition and mutualism is

D

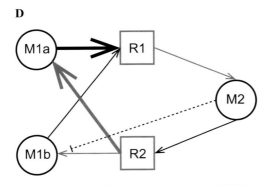

Figure 1. Competition for mutualistic resources. (A) The traditional net effects diagram of mutualism. The arrows show the net reciprocal benefits (+/+) of the interaction between two mutualistic species (M1 and M2). (B) A resource-based diagram

of mutualism. Mutualists M1 and M2 produce resources R1 and R2, respectively, and consume the resource produced by the partner mutualist. The arrows show resource production (black) and resource consumption (red). (C) Mutualism when low-quality mutualists have a competitive advantage. M1 has now been decomposed into two individuals or two species from a mutualist guild (M1a and M1b). M1a is a better mutualist since it offers a large amount of resources to M2 (thick arrow from M1a to R1). However, M1b is a better competitor for the resources produced by M2 (thick arrow from R2 to M1b). Here, the competitive advantage comes from interference competition (dashed inhibition arrow between M1b and M1a's R2 consumption arrow). (D) Mutualism when high-quality mutualists have a competitive advantage. Again, M1a is a better mutualist by producing large amounts of R1. However, it is now a better competitor for R2 (thick arrow from R2 to M1a). In this case, the competitive advantage results from partner control by M2 (dashed inhibition arrow between M2 and M1b's R2 consumption arrow).

that there is competition for the commodities mutualists produce. Consider pollination, the most thoroughly studied mutualism. Floral visitors are well documented to compete intraspecifically for nectar, for example, by adjusting their foraging routes in relation to the activities of conspecifics on flowers.[17,18] Competition may also pit individuals of different species against one another, a phenomenon particularly well-investigated between honeybees and native bees that they may be displacing.[19] On the other side of the interaction, it is common for plants to fiercely compete both intraspecifically and interspecifically for pollinators, as Mitchell *et al.* have thoroughly reviewed;[20] competition for pollination has clearly shaped the evolution of flower sizes and numbers, floral reward chemistry and volume, and both visual and olfactory cues.[20–23]

Adding competition for resources explicitly into Figure 1B yields many possible outcomes, two of which are shown in Figure 1C and D. First, we decompose one mutualist (M1) into multiple individuals or species; we will explore how these entities (e.g., different pollinator individuals or species, or different nectar-producing plant individuals or species) compete for mutualistic resources. We will call these M1a and M1b. These two entities consume a shared resource, R2, and both produce a second resource, R1. Via these resources, both interact with a partner species, M2. However, M1a and M1b are not identical. The different thicknesses of the lines from M1a and M1b to R1 indicate that one entity produces more of R1 than does the other. Similarly, the different thicknesses of the lines from R2 to M1a and M1b indicate that one entity consumes more of R2 than does the other.

Figure 1C illustrates the situation in which a superior competitor is an inferior mutualist and competitive interactions play out in a way likely to be harmful to the shared partner. In Figure 1C, M1b provides less of R1 to the shared partner M2; thus, we define it as an inferior mutualist. However, M1b uses more of R2 than does the superior mutualist, M1a. M1a's lower consumption of R2 is due in some way to the presence of M1b: M1b might be actively interfering with M1a's consumption, or it may simply be consuming R2 faster or more efficiently, leaving less behind. We illustrate this effect with a dashed inhibition arrow running from the superior competitor M1b to the arrow connecting the shared resource R2 to M1a. To put this in words, the shared partner is

stuck with a relatively low-quality mutualist able to reduce the success of better-quality mutualists. (This is quite realistic biologically. For example, Bennett and Bever demonstrated that the most beneficial mycorrhizal fungus species for *Plantago lanceolata* is the worst competitor for root space, whereas the worst fungal mutualist is the best competitor for *P. lanceolata* roots.[9]) This is a situation that might lead to the low-quality mutualist dropping resource provision altogether and becoming an exploiter of the system, leading one to question how mutualisms embodying this structure are able to persist evolutionarily. We discuss this in more detail below.

In Figure 1D, we illustrate the situation in which it is the superior mutualist that holds the competitive advantage. In contrast to the situation shown in Figure 1C, this can clearly benefit the shared partner. Here, M1a, the mutualist that consumes more of the shared resource R2, also provides more of resource R1 to the partner species M2. We have illustrated this outcome as being the result of actions of M2: it has exerted some kind of "partner control" that reduces the ability of the inferior mutualist M1b to compete for the resources it provides. This situation has been argued to permit mutualisms to persist in the presence of cheaters, as we discuss below. As in the case shown in Figure 1C, it is not difficult to identify biological examples of these relationships. Adam documented such a case in the interaction between cleaner fish and their "clients."[10] Clients (butterflyfish) are able to selectively associate with cleaners (wrasses) that provide them with the highest quality service (parasite removal). Thus, clients confer a competitive advantage to the best cleaners. Indeed, cleaners provide better service when competitors are present as this is the only way that they will be chosen by hosts.

Figure 1C and D are only two possible types of competitive interactions within guilds of interacting mutualists. Many other networks can be envisioned, and indeed are well documented in the literature. For example, we have not considered here that competitive advantage and mutualistic quality can both be functions of population size, and thus can vary over ecological time scales (e.g., Refs. 24 and 25). There is clearly much more to explore. Our overall point is simply that making the resource exchange underlying mutualism explicit (Fig. 1B), and clarifying which of these resources are competed for, which partners hold the competitive advantage, and which are the best mutualists (Fig. 1C and D), reveal a fascinating range of possible ecological and evolutionary ramifications that are completely obscured in the simple, standard net effects-based way of viewing mutualisms (Fig. 1A).

Competition between mutualists and exploiters

Almost all mutualisms are afflicted with individuals and species that gain the benefits that mutualism offers, while investing little or nothing in return.[26–28] A perennial question about mutualism is how it can persist ecologically and evolutionarily in the face of these organisms (hereafter, *exploiters*) that would seem to be at an advantage. To answer this question, it is essential to think beyond the comparative effects of exploiters and mutualists on their shared partners. These are relatively well studied. One also needs to consider the nature and outcome of competition between exploiters and the species that share that partner. This issue has barely been addressed in the growing literature on cheating within mutualism (but see, for instance, Refs. 13 and 29–31).

First, it is necessary to illustrate the interactions in question, as we did in Figure 1, for mutualisms in the absence of exploitation. Parallel to Figure 1A, Figure 2A gives the standard, net effects-based illustration of a mutualism that is associated with an exploiter, E. The arrows are labeled to indicate that E benefits from M1 but is detrimental to it, and that E and M2 are detrimental to each other (since they share a partner). A well-known example is the well-studied network of interactions among plants, pollinators, and nectar-robbers, floral visitors that feed on nectar but do not pick up or deposit pollen.[32] Nectar-robbers (E) and pollinators (M2) both interact with plants (M1), but only M2 confers a benefit to M1. As in Figure 1A, no mechanisms are shown.

The other panels in Figure 2 take a resource-exchange rather than a net effects-based approach to these interactions, as did Figure 1B–D. The resources of exchange are added into Figure 2B, parallel to Figure 1B. Note that the M1-R1-M2-R2 network in Figure 2B is identical to that shown in Figure 1B. However, an exploiter species has now been added. Like M2, E consumes the resource R1, but unlike M2, it does not provide the resource R2 to the partner M1. Interestingly, moving to a resource-exchange perspective has served to simplify the net

effects depiction (Fig. 2A) by making it clear what it is that species share, exchange, and compete for. In the case of plant—pollinator–nectar-robber interactions, for example, Figure 2B clarifies that pollinators (M2) and nectar-robbers (E) both utilize a resource, nectar (R1), but that only pollinators deliver a resource (R2), pollen transport, to the shared partner.

Parallel to Figure 1C and D, Figure 2C–E shows distinct ways in which competition for a resource shared between a mutualist and exploiter can occur and be mediated. Each suggests distinct ecological and evolutionary outcomes, and each captures a phenomenon represented in the empirical literature.

In Figure 2C, the exploiter is competitively superior at obtaining the resource from the shared mutualist. (We refer readers to the discussion of Figure 1C for an explanation of how to read these effects based on the colors, arrow thicknesses, and arrow patterns.) This gives rise to a situation in which mutualists can potentially be competitively excluded by exploiters, raising the obvious question of if and how mutualism can persist under these conditions. Examples can be found in nature. For example, Dohzono *et al.* studied a case in which a nectar-robbing bumble bee, *Bombus terrestris*, competes with native pollinating bumble bees for a shared nectar-producing plant, *Corydalis ambigua*, in Japan.[33] Once nectar-robbers are present, pollinators abandon *C. ambigua* for other nectar resources, to the detriment of the plant. It is interesting that in this case, the exploiter is an introduced species. Will this plant, or at least its mutualism with native pollinators, be able to persist over the long term? This is more than an abstract evolutionary question. It highlights that an understanding of competitive hierarchies among native and introduced species may shed light on the conditions under which mutualisms will be able to persist and evolve in the face of anthropogenic change.[34]

Figure 2C may thus seem to be an ecologically and evolutionarily fragile situation for mutualism. However, close study of several mutualist–exploiter interactions has revealed that the shared partner has some ability to control the exploiter in a way that shifts the competitive advantage toward the mutualistic partner. We illustrate this general situation in Figure 2D. As a biological example, Kiers and colleagues have elegantly demonstrated that plants are

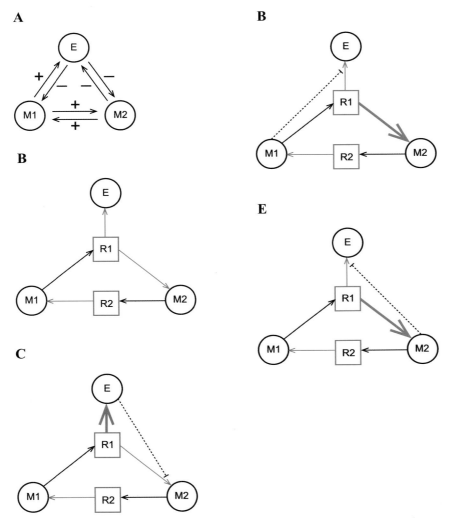

Figure 2. Competition between mutualists and exploiters. (A) A net effects diagram. The core mutualism between M1 and M2 is the same as in Figure 1A. An exploiter (E) has been added. E gains a benefit (+) from M1, but inflicts a net cost (−) on M1, since E does not reciprocate any benefit. Both E and M2 gain benefits from M1, thus they are competitors (−/−) for these benefits. (B) A resource-based diagram. The core mutualism is the same as in Figure 1B. The added species, E, does not produce any resources. However, E consumes R1 thereby competing with M2 for this resource. (C) Exploited mutualism when the exploiter has a competitive advantage. E is a better competitor for R1. Here, the competitive advantage comes from interference competition. (D) Exploited mutualism when partner control gives a competitive advantage to the mutualistic partner. M2 is able to gain more of R1 due to partner control by M1. (E) Exploited mutualism when mutualists are superior competitors. Again, M2 is able to gain more of R1; however, it is now because M2 interferes directly with E. For the resource-based diagrams, (B–E), arrows represent resource production (black arrows), resource consumption (red arrows), increased resource production or consumption (thick arrows), and interference with resource consumption (dashed inhibition arrows).

able to discriminate among and differentially deliver resources to symbiotic *Rhizobium* bacteria that produce relatively more fixed nitrogen for them;[35,36] similar control mechanisms may exist within other plant rhizosphere mutualisms.[37] A wide variety of control mechanisms have been suggested under various names (e.g., sanctions, punishment, and partner choice). Given the potential importance of these mechanisms in allowing mutualism to persist in the face of exploitation, a large body of theory has been developed to examine when each mechanism is likely to evolve and how it would function (e.g.,

Refs. 27 and 38–40). We consider this issue in more depth below.

It is clear, however, that partner control mechanisms are not always necessary to explain how competing mutualists and exploiters are able to coexist. One obvious case is when it is the mutualist rather than the exploiter that holds an innate competitive advantage. This situation is illustrated in Figure 2E. Good empirical evidence comes from mutualisms between certain tropical plants and the highly specialized ants that inhabit them. Some of these ants are mutualistic, aggressively defending their plants from herbivore attack, whereas others occupy the plant and provide no defense. Exploiter ants have in several cases been reported to be competitively inferior to mutualistic ants; when mutualists invade a plant that exploiters occupy, it is commonly the exploiters that are displaced.[41,42] Their superior ability to locate unoccupied plants results in a competition–colonization trade-off that allows them to persist even in the face of their evident disadvantage when challenged.[41,42] Furthermore, certain mutualistic ants possess dietary specializations that allow them to more efficiently use the food that plants provide them.[29,31] Of course, it is possible that the chemical makeup of this resource has evolved as a partner-control mechanism that shifts the competitive balance toward mutualists, illustrating that it can be difficult to empirically distinguish interaction networks as similar as those shown in Figure 2D and E.

Thus, to understand when we would expect to see the evolution of sanctions and punishments in mutualisms afflicted with exploitation, it is critical to examine whether exploiters hold a competitive advantage over the mutualists with which they share a resource, or vice versa. This has only rarely been investigated. The usual assumption has been that control mechanisms are always essential for mutualism to persist in the face of exploitation.

Competition between mutualists

In the previous two sections, we have considered the two best understood ways in which competition and mutualism interact: when individuals or species compete for resources provided by a shared mutualist and when mutualists and exploiters compete for resources from a shared partner. A much less studied phenomenon is when mutualistic partners are also competitors for resources. When mutualists

occupy different trophic levels, as is generally the case,[2] resource sharing is not expected. Thus, for instance, plants and pollinators do not share and compete for resources, nor do ants and the plants they defend. However, a number of less well-known mutualisms involve species that occupy the same trophic level and thus are likely to compete for access to shared resources.[43] Müllerian mimicry in butterflies provides a good illustration. In these cases, mutual benefit is derived by a shared resemblance that "trains" predators to recognize them and, because they are distasteful or toxic, to avoid consuming them (e.g., Ref. 44). These individuals are likely to compete for food and other resources, however. A similar situation can be found in interspecific group foraging: individuals benefit either by shared predator vigilance or by increased access to food.[45,46] However, they also may compete for the food they locate.[47]

We know of no standard way of illustrating these relationships via net effects arrows as in Figures 1A and 2A. We attempt to do this in Figure 3A. Here, the arrows are labeled +/– because the partners are simultaneously competing (hence, are in a minus/minus interaction) and benefiting each other (hence, are in a plus/plus interaction). This is not very satisfying. The difficulty of finding a way to illustrate these interactions is symptomatic of a general problem with using net effects to capture them: mutualism and competition occur simultaneously, and whether they add up to net effects that are positive for one, both, or neither partner is likely dependent on many system-specific and context-dependent factors.[48] Path coefficients could be used to quantify the net effects;[16] however, the individual contributions of mutualism and competition would be lost, making it difficult to translate the relationship into a mechanistic model.

These interactions are much more effectively captured once resources are illustrated explicitly, as seen in Figure 3B. Now it is clear that M1 and M2 interact mutualistically via resources R1 and R2, exactly as in Figures 1B and 2B. What is different here is that there is a third resource, R3, for which M1 and M2 compete. To frame the group foraging example described above in these terms, investment of time and energy into predator vigilance might be the resource of exchange. Indeed, the resources R1 and R2 are in this case the same thing. R3 in this example might be a shared food resource.

A

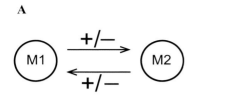

B

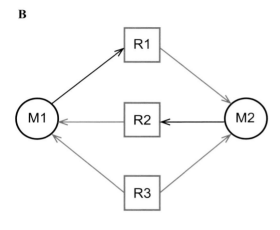

C

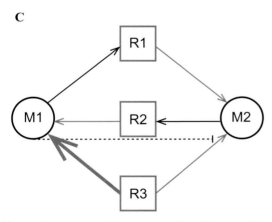

Figure 3. Competition between mutualists. (A) A net effects diagram. M1 and M2 are simultaneously engaged in mutualism (+/+) and competition (−/−). The true net effects could be either positive or negative, depending on whether mutualism or competition dominates, respectively. (B) A resource-based diagram. The core mutualism is the same as in Figure 1B. There is now also a third resource (R3) that is consumed by both M1 and M2. (C) Asymmetric competition between mutualists. M1 is a better competitor for the shared resource and interferes with consumption of R3 by M2. While not explicitly shown, competition for R3 could change production of the mutualistic resources, R1

The central question for understanding the persistence of mutualism in this scenario is when the magnitude of competition will outweigh the magnitude of mutualism or vice versa, and, if competition is stronger, what the fate of mutualism is likely to be. Clearly, we cannot project the evolutionary consequences for mutualism in an interaction network like this without explicitly considering and measuring competition.

Figure 3C illustrates one way in which competition could be manifest. In this case, M1 is the superior competitor for the shared resource, suppressing M2's use of it. When framed this way, the question about the persistence of mutualism becomes refocused as a question about the persistence of M2. Will the detriment M2 experiences from its reduced access to R3 outweigh the benefit it receives from M1, via the mutualistic component of their interaction? And, how will these combined effects feed back on M1? Competition could logically lead loose mutualisms of this type to dissolve as predicted in a model of Ranta *et al.*[48]

Relevant empirical data on these questions are few. Hino found that five of six bird species studied changed their foraging behavior when in mixed-species flocks compared to when foraging alone.[49] Interestingly, feeding rates were higher in mixed flocks. Although this may indicate an absence of competition for food, and in fact an increase in food availability, the authors point out that it could also be an effect of kleptoparasitism or social learning, either of which could be the result of intense competition among species.

As in all previous cases we have described, our figures capture some but not all of the complexity of how competition and mutualism can interact. In Figure 3B, competition is for a resource extrinsic to mutualism (R3). However, competition may also occur between M1 and M2 for mutualistic resources (R1 and R2). For example, two studies in marine habitats have found that two fish species collaborate to locate food, but then appear to compete to consume it.[47,50] To capture these and other complex competition–mutualism interactions (e.g.,

and R2. For the resource-based diagrams, (B, C), arrows represent resource production (black arrows), resource consumption (red arrows), increased resource consumption (thick arrow), and interference with resource consumption (dashed inhibition arrow).

Ref. 51), some important modifications to our figures would be required. The ones we show, however, provide a starting point for how to conceptualize these phenomena.

Competitive advantage as a benefit of mutualism

The best studied benefits of mutualism are transportation (e.g., of pollen by pollinators), protection (e.g., of aphids by ants), and nutrition (e.g., of plants by *Rhizobium* bacteria). However, other mutualistic benefits are well documented. Among these are beneficial alterations of a partner's competitive environment, and this is the final intersection of mutualism and competition that we will consider. Here are two empirical examples. Hartnett *et al.* explored how competitive interactions among prairie plants might be mediated by mutualistic mycorrhizal fungi.[52] They demonstrated that competitive dominance of one grass species, *Andropogon gerardii*, depends on it having access to mycorrhizae. Thus, in this case, a mutualist confers traits that give its partner a competitive advantage. As another example, Stachowicz and Hay studied interactions between herbivorous crabs and the coralline algae upon which they live.[53] They showed that crabs feed upon fouling seaweeds that, if unchecked, would overgrow the coralline algae; the algae provide a place for crabs to live. In this case, then, a mutualist actively interferes with a competitor to the partner's advantage.

The net effects figure shown in Figure 4A summarizes interactions of this general type. Note that in this case, there might or might not be a mutualism between M1 and M2 in the absence of the competitor C. With reference to the two examples above, plants and mycorrhizae are likely to be mutualists even in the absence of competitors as there are other benefits of this interaction. However, the crabs and algae studied by Stachowicz and Hay would likely not be.[53] Such context dependency—that is, a mutualistic outcome that occurs only in a limited set of environments—is almost impossible to capture in a net effects-based figure such as Figure 4A. A resource-based figure permits us to do this. Furthermore, it allows us to recognize important differences between phenomena exemplified in these two empirical cases and to consider how competition may function in each of them. For this reason, we treat them separately below.

Figures 4B and C illustrate the case in which one, but not the only, benefit of mutualism is the suppression of competitors. In Figure 4B, we again see the same core mutualism as in Figures 1B, 2B, and 3B. Added to it is a competitor (C) that shares a different resource (R3) with mutualist M1. (If C shared R1 or R2 with a mutualist, we would consider it to be an exploiter of the M1–M2 mutualism, and the scheme shown in Fig. 2B would be more appropriate.) As in Figures 1B, 2B, and 3B, mutualist M1 gains a direct benefit from consuming R2. In addition, as illustrated in Figure 4C, consuming R2 gives M1 a competitive advantage over C for the shared resource R3. The example provided by Hartnett *et al.*,[52] described above, fits this scenario. Note again that it is not M2 (mycorrhizae) that suppress the competitor; M1 (the plant) does this, but only when their mutualists M2 are present. Competitive suppression is thus an indirect benefit provided by mycorrhizae accompanying the direct benefits of nutrient provision. Evidently, it can be extremely important in its own right, however. For example, Wilson and Hartnett show that community-scale plant diversity may be increased if mycorrhizae augment a subordinate species' performance in competition with a dominant one.[54]

Figure 4D illustrates the case when an interaction is mutualistic *only* in the presence of a competitor, that is, when alteration of the competitive environment is the only benefit of a mutualism. In this case, there is no core mutualism resembling those in Figures 1B, 2B, and 3B. Here, one species (M2) consumes a resource (R1) provided by a partner (M1), but there is no reciprocal benefit (i.e., there is no resource R2). As in Figure 4B, the resource-providing partner M1 competes for another resource (R3) with a competitor C. In Figure 4D, the relationship between M1 and M2 is not mutualistic: these interactions are often referred to as commensal (i.e., +0 rather than ++) or facilitative, or sometimes more generically as "positive interactions" (e.g., Refs. 55 and 56). Alternatively, if R1 is costly to produce and no benefit is returned for its provision, then this interaction could be antagonstic (+−). It can become mutualistic, however, via the mechanism illustrated in Figure 4E. Here, M2 alters the environment in a way that shifts the balance of competition between M1 and C in favor of M1. This scenario matches that described by Stachowicz and Hay,[53] in which crabs benefit from

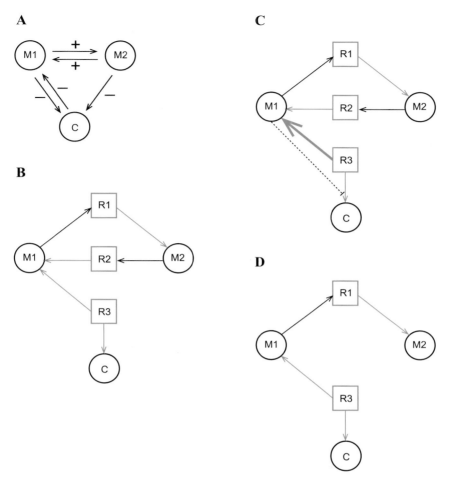

Figure 4. Competitive advantage as a benefit of mutualism. (A) A net effects diagram. The core mutualism is the same as in Figure 1A. A third species (C) has been added that competes (–/–) with M1. M2 interferes (–) with C, thus giving an indirect benefit to M1, in addition to any direct benefits of the core mutualism. (B) A resource-based diagram. The core mutualism is the same as in Figure 1B. Additionally, M1 consumes a resource (R3) that is also consumed by C. (C) Competitive advantage is a secondary benefit of mutualism. By consuming R2, M1 becomes a superior competitor for R3 and is able to interfere with consumption of R3 by C. (D) A resource-based diagram of context-dependent mutualism. M1 produces R1, which is consumed by M2. However, M2 does not produce any resource that can be consumed directly by M1 (R2 has been removed from the core mutualism). Instead, the benefit provided by M2 is context-dependent and requires the presence of C. (E) Competitive advantage is the only benefit of mutualism. M2 is an antagonist of C and interferes with its consumption of R3. Consequently, M1 is able to increase consumption of R3. For the resource-based diagrams, (B–E), arrows represent resource production (black arrows), resource consumption (red arrows), increased resource production or consumption (thick arrows), and interference with resource consumption (dashed inhibition arrows).

coralline algae (they are provided with a substrate on which to live and feed), but algae only benefit from the crabs when fouling seaweeds are present and crabs remove them.

Context-dependent outcomes like this one are now widely recognized as one of the most ubiquitous ecological features of mutualism, regardless of their natural history.[3] The evolutionary implications of context dependency, in contrast, have barely been considered (but see Ref. 57). Figure 4 clearly suggests that understanding when mutualisms and mutualistic outcomes arise may depend upon documenting the competitive environment in which they occur.

E

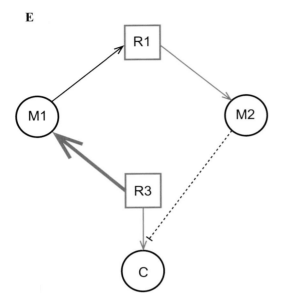

Figure 4. *Continued*

Mechanisms of competitive asymmetry among mutualists: the cooperation–competition trade-off

In the previous section, we showed that competition for resources is a common feature of mutualisms. Additionally, we discussed how competition is often between individuals or species that vary in their quality as mutualist partners (Fig. 1C and D), and in the extreme case are exploiters that do not reciprocate any benefits (Fig. 2). This variation in partner quality raises fundamental evolutionary questions: how is the variation in partner quality maintained and how can mutualisms persist despite low-quality partners and exploiters? If low-quality mutualists and exploiters have a competitive advantage, evolution should lead to the loss of mutualism. Thus, the question becomes: what mechanisms favor high-quality mutualists? Below, we demonstrate how considering the details of competition described above is crucial for understanding how investment in mutualism is maintained evolutionarily.

Although evidence is lacking, it has generally been assumed that individuals can gain a competitive advantage by investing less in rewarding partners (e.g., Figs. 1C, 2C, and 5A).[58,59] In this case, or even when individuals that contribute less to the partner are *equal* in direct competition, the mutualism is expected to evolve to extinction.[11] This occurs as a result of a "tragedy of the commons"[60]: lower-quality mutualists can invest more in survival and reproduction while retaining full benefits from partners, with an ever-growing numerical advantage as a consequence.

Evolutionary models of mutualism have invoked several mechanisms that shift the balance in favor of higher-quality mutualists. What these mechanisms share in common is that they enable selection for increased investment in mutualism by offsetting the cost of investment with the benefit of a competitive advantage. How the competitive advantage arises depends on the mechanism. We discuss the most prominent of these mechanisms below, along with empirical examples. For each mechanism, we describe how it functions and how it affects competition. As we shall see, the cooperation–competition trade-off shaped by these mechanisms is key to understanding several important aspects of the evolution of mutualisms.

Competitive asymmetry caused by partner choice

Interactions with inferior partners can be costly and reduce the potential for interactions with better partners. Mutualists can escape these costs by using "partner choice" mechanisms to restrict interactions to the best available partners.[38,61] In turn, partner choice alters the competitive and adaptive landscape by conferring an advantage to better mutualists (e.g., Figs. 1D, 2D, and 5B).

Partner choice mechanisms may be based on a comparison of available partners. This "active partner choice" is common for animals such as pollinators and fruit dispersers that can choose partners based on reliable cues of reward quality,[16] and is also well-described in clients of cleaner fish.[62] Since high-quality partners are given preferential access to rewards their mutualists provide, they gain a competitive advantage over any lower-quality partners.

Alternatively, partner choice mechanisms may be based on static criteria that prospective partners must meet. Examples include legumes whose roots can only be nodulated by specific nitrogen-fixing rhizobia through a "lock-and-key" recognition system,[63] and bobtail squid whose light organs can only be detoxified and colonized by bioluminescent *Vibrio fischeri* bacteria.[64] In this case, only individuals that meet the choice criteria can successfully compete for mutualist rewards.

trade-offs between competitive dominance and other traits, such as fecundity, can select for partners that provide a higher net benefit.[89,111] Thus, details of the sanction process, its relation with competition, and trade-offs with other traits, are critical for the evolution of mutualist quality.

In general, the evolutionary implications of partner control for mutualist quality depend on the relationship between mutualist quality and competitive ability (Fig. 5). The parallel evolution of large investments by mutualist partners is expected if competition for partners is strongly asymmetrical in both mutualists, that is, if the competitive advantage to better mutualists is high.[11] This means that there is a "cooperation–competition" trade-off between the direct cost of mutualistic investment and the direct benefit of competitive advantage (Fig. 5); and that the slope of the trade-off is steep. This prediction is robust to the inclusion of (nonevolving) exploiters.[13] However, mutualist–exploiter coevolution tends to degrade the mutualist's quality.[111] A clear-cut prediction is that mutualist and exploiter converge, resulting in relatively benign exploitation, if mutualists experience strong competition for partners; they diverge, leading to severe exploitation, if competition for partners is weak.[111]

Evolutionary patterns of convergence and divergence

Mutualists' traits may be expected to coevolve to match in order to maximize the exchange of rewards (e.g., Refs. 120 and 121, but see Ref. 122), and convergent evolution has been found between species that share mutualist partners.[123] Furthermore, there can be "advergent" evolution, such as when exploiters (e.g., nectarless orchids) evolve mimicry of a mutualist species.[124]

Convergence may occur between other aspects of the mutualists' niches and this is particularly important for understanding mutualism between resource competitors (Fig. 3). In these cases, mutualism is expected to promote coexistence despite overlapping niches, as the benefits of mutualism can compensate for (some of) the costs of competition.[125,126] However, competition should still constrain the evolution of mutualism since increasing the partner population decreases resources available.[127]

In the case of Müllerian mimicry, the benefits of mutualism are enhanced if the mimics overlap

in time and space. However, this overlap is likely to increase competition for shared resources. Either mutualism or competition may be the dominant interaction. In neotropical butterflies, mimicry has driven niche convergence between sympatric, unrelated species.[128] In contrast, mimicry in catfish occurs almost exclusively between species that are not competitors.[129] More research is needed to determine how often and why mutualism is dominant to competition and vice versa.

Divergence between mutualists can be driven by competition for partners. Trade-offs in the ability to exploit different shared resources can select for resource specialization;[130] similarly, competition for partners can select for specialization on a subset of potential partner species.[15,102] For example, the diversification of floral morphology and phenology within communities has been attributed to competition for partners.[131–133] Interestingly, the intensity of competition for partners can be reflected in the degree of diversification; plants competing more for pollinators than for fruit dispersers show higher diversification in flower than in fruit morphology.[134] Nevertheless, it has been suggested that, even without specialization, mutualists competing for a shared partner (Fig. 1C and D) should be more likely to coexist than general competitors, since the competing mutualists can increase the density of their shared resource (the partner species).[14]

Competition can be intense even within specialized interactions, especially when one mutualist partner utilizes competition in partner control. As described above, mutualists may evolve mechanisms that increase the intensity of competition among their partners.[89,90] Selection should then act on the partner species to reduce competition. As alternatives to evolving to become better competitors, partners could diversify in their use of shared mutualist commodities,[89,135] or abandon the mutualism.[58] Even when there are mechanisms that give a competitive advantage to mutualists, intense competition among mutualists has been predicted to enable the coexistence of exploiters,[25] as well as to make conditions favorable for exploiters to arise from mutualists.[11,89,111,135] If the competitive advantage does not completely compensate for the costs of investment in mutualistic resources, the mutualism may remain susceptible to invasion by low-quality partners.[136] Some empirical support for these predictions comes from the pollination/seed-predation mutualisms of globeflowers and yuccas.[137–141] In these mutualisms, the pollinators have responded to larval competition by diversifying in timing and location of oviposition, with the later visitors frequently becoming exploiters of the mutualism.

The long-term evolutionary implications of competition for partners critically depend on the shape of the cooperation–competition trade-off.[11] One possibility is a *rewarding* trade-off (Fig. 5B), in which the competitive advantage of being a better mutualist is large while the competitive disadvantage suffered for being an inferior mutualist is small. Sufficiently rewarding trade-offs can result in evolutionary diversification of mutualist quality and maintenance of variation in mutualist quality. In contrast, the trade-off could be *punishing* (Fig. 5B), in which increasing mutualist quality adds little to competitive ability, whereas there is a large loss of competitive ability with a small decrease in mutualist quality. Punishing trade-offs oppose evolutionary diversification.

We still have only a preliminary understanding of how the cooperation–competition trade-off is shaped by partner control mechanisms, and in particular whether partner choice versus sanctions result in qualitatively different shapes. The difference between rewarding and punishing shapes might not be between partner choice and sanctions, but rather between partner control mechanisms that directly compare partners versus those that compare partners to fixed criteria. For direct comparisons, the advantage comes from being superior to the rest of the potential partners. A small increase in quality in a low-quality mutualist is unlikely to increase its chances of being the best of the available partners. However, the same small increase in quality might move a high-quality mutualist up to the "best partner" rank—thus leading to a rewarding trade-off. On the other hand, if the partners are compared to fixed criteria, all that is necessary is to meet the criteria, and any further increases in mutualist quality should not have much effect on competitive advantage—thus leading to a punishing trade-off. Increases in quality for high-quality mutualists may even lead to a net decrease in competitive ability if the cost of investing in mutualism is taken into account (a concave trade-off, see Fig. 5C). Empirical data are now needed to back up these heuristics.

Conclusions

Nearly all mutualisms have consumer–resource interactions embedded within them, whereby species on one side of the interaction exploit, and therefore may compete for, a resource (food, service) produced by species on the other side.[2] A natural consequence of this principle is that mutualisms set the stage for competitive interactions. As we showed in the section. Competition and mutualism: empirical phenomena, mutualists compete as consumers of their partner-produced resources; the scope of competition may be even broader if, for example, the mutualistic resource influences the outcome of competition for other resources. This leads us to advocate for the importance of competition, both intra- and interspecific, as a critical factor of the ecological and evolutionary dynamics of mutualisms. More generally, to understand mutualism, it is critical to quantify the competitive interactions that lie at its heart.

Our review has highlighted the various form of competition that have been well documented to be associated with mutualistic interactions. We have offered a simple but general classification to unite similar phenomena and distinguish among divergent ones. As we have shown, based on existing theory, the intensity of intra- and interspecific competition for partners, as well as asymmetries in competitive effects on individuals that differ in their mutualistic quality, profoundly affects the ecological stability and evolutionary dynamics of mutualism. Yet, our understanding of the intersection between mutualism and competition is in its infancy. Further studies are necessary to illuminate:

- how multiple effects of competition combine to influence the dynamics of the interaction;
- how, conversely, the nature of the traded resources/services influences, both ecologically and evolutionarily, the type and intensity of competition and the underlying mechanisms; and
- how the structure and diversity of a mutualistic network and the competitive interactions embedded in it may jointly coevolve and influence the short-term persistence of mutualists, the long-term stability of their interaction, and the function of the network in its broader ecosystem context.

To address these broad issues at the interface between competition and mutualism, we advocate a joint empirical–theoretical approach. On the theoretical side, we need to extend modeling approaches that integrate ecological and evolutionary processes, for example, adaptive dynamics and related models.[142,143] At the level of individuals, key elements of these models will be the existence and shape of physiological trade-offs between cooperation (i.e., investment in mutualism), competition, and other functional traits; the capacity for individuals to express conditional responses to variation in their mutualistic and competitive environment; the existence of heritable variation for mutualistic and competitive traits and their potentially conditional expression; and the structure of the network of other consumer–resource interactions in which individuals are embedded. Eco-evolutionary models should be designed that integrate knowledge about these individual-level properties and that scale up to gain insights into their ecological and evolutionary consequences. It is thus critical that we deepen such knowledge on the empirical side. To this end, model systems are needed in which multiple forms of competition interact with mutualism.

For a given system, the empirical agenda must start by asking (i) whether partners on at least one side of the interaction compete for mutualistic resources (Fig. 1B); (ii) whether exploiters are present (Fig. 2B), and if so, how costly they are; (iii) whether mutualists also compete for some shared resource (Fig. 3B); and (iv) whether gaining access to mutualists provides one partner with an advantage in competition with its guild members (Fig. 4B).

Observations and manipulative experiments should be performed to weigh, for instance, the strength of competitive effects relative to mutualistic effects and to examine evidence for adaptations that reduce or increase the intensity of competition. The best studied mutualisms from the perspective of the questions raised here are plant—pollinator–nectar-robber and plant—ant—defender—ant–opportunist systems. However, for no system have all four questions been answered conclusively. For example, evidence that competition between pollinators and nectar-robbers inflicts a fitness cost on plants is equivocal.[32] A review by Bergstrom *et al.* provides a useful basis to discuss how other mutualisms position themselves with respect to these four axes of variation and how they could serve the objectives of a joint empirical and theoretical research program on the interplay between mutualism and competition.[144]

The competition perspective on mutualism opens interesting approaches for further unification of the ecological and evolutionary analysis of interspecific interactions. Particularly promising to explore are parallels with the ecology and evolutionary biology of host–parasite systems. Interactions between hosts and parasites have long been studied as consumer–resource interactions, and the role that competition plays in their ecology and evolution has already received considerable attention. Given the strong conceptual and biological ties that exist between mutualism and parasitism, we anticipate fruitful fertilization of the nascent studies of competition in mutualism from the ongoing theoretical and empirical investigation of competition in parasitism. Taking a competition perspective to parasitism was key to discovering the diverse implications of multiple infection by mixed strains of parasites for virulence.[145] This line of inquiry began with simple mathematical models showing how influential within-host competition can be for the evolution of virulence.[146–149] This spawned further theoretical and empirical studies that unraveled considerable variation in the relationship between virulence and competitive ability,[150–154] and highlighted the complexity of real competitive interactions among parasites for host resources compared to simple resource competition.[155] Another exciting parallel between parasitism and mutualism in the context of competition can be drawn from the issue of specialization and diversification. Much theory has been developed to shed light on the evolution of host specificity in parasites and the consequences of specialization for ecological coexistence and evolutionary diversification. Similarly, we expect theoretical advances to shed light on the evolution of mutualistic networks by considering feedbacks with the ecological mixture of trade and competition among partners. This work will help to strengthen our understanding of the structure and diversity of mutualistic networks, as well as to foster our ability to predict their robustness to environmental perturbation.

Acknowledgments

We thank M. Friesen and two anonymous reviewers for helpful comments on the manuscript. E.I.J. was supported by National Science Foundation Grant DMS-0540524 to R. Gomulkiewicz. R.F. acknowledges support from National Science Foundation Award EF-0623632, the Institut Universitaire de France, and the Agence Nationale de la Recherche.

Conflicts of interest

The authors declare no conflicts of interest.

References

1. Bronstein, J.L., R. Alarcón & M. Geber. 2006. Tansley review: evolution of insect/plant mutualisms. *New Phytol.* **172:** 412–428.
2. Holland, J.N., J.H. Ness, A. Boyle & J.L. Bronstein. 2005. Mutualisms as consumer-resource interactions. In *Ecology of Predator-Prey Interactions*. P. Barbosa, Ed. Oxford University Press. Oxford.
3. Bronstein, J.L. 2009. Mutualism and symbiosis. In *Princeton Guide to Ecology*. S.A. Levin, Ed. Princeton University Press. Princeton.
4. Holland, J.N. & D.L. DeAngelis. 2010. A consumer-resource approach to the density-dependent population dynamics of mutualism. *Ecology* **91:** 1286–1295.
5. Dworschak, K. & N. Blüthgen. 2010. Networks and dominance hierarchies: does interspecific aggression explain flower partitioning among stingless bees? *Ecol. Entomol.* **35:** 216–225.
6. Hegland, S.J., J.A. Grytnes & O. Totland. 2009. The relative importance of positive and negative interactions for pollinator attraction in a plant community. *Ecol. Res.* **24:** 929–936.
7. Ness, J.H., W.F. Morris & J.L. Bronstein. 2006. Integrating quality and quantity of mutualistic service to contrast ant species visiting *Ferocactus wislizeni*, a plant with extrafloral nectaries. *Ecology* **87:** 912–921.
8. Rodríguez-Gironés, M.A. & L. Santamaria. 2007. Resource competition, character displacement, and the evolution of deep corolla tubes. *Am. Nat.* **170:** 455–464.
9. Bennett, A.E. & J.D. Bever. 2009. Trade-offs between arbuscular mycorrhizal fungal competitive ability and host growth promotion in *Plantago lanceolata*. *Oecologia* **160:** 807–816.
10. Adam, T.C. 2010. Competition encourages cooperation: client fish receive higher-quality service when cleaner fish compete. *Anim. Behav.* **79:** 1183–1189.
11. Ferrière, R., J.L. Bronstein, S. Rinaldi, R. Law & M. Gauduchon. 2002. Cheating and the evolutionary stability of mutualisms. *Proc. R. Soc. Lond. Ser. B-Biol. Sci.* **269:** 773–780.
12. Abrams, P.A. & M. Nakajima. 2007. Does competition between resources change the competition between their consumers to mutualism? Variations on two themes by Vandermeer. *Am. Nat.* **170:** 744–757.
13. Ferrière, R., M. Gauduchon & J.L. Bronstein. 2007. Evolution and persistence of obligate mutualists and exploiters: competition for partners and evolutionary immunization. *Ecol. Lett.* **10:** 115–126.
14. Lee, C.T. & B.D. Inouye. 2010. Mutualism between consumers and their shared resource can promote competitive coexistence. *Am. Nat.* **175:** 277–288.
15. Palmer, T.M., M.L. Stanton & T.P. Young. 2003. Competition and coexistence: exploring mechanisms that restrict and maintain diversity within mutualist guilds. *Am. Nat.* **162:** S63–S79.
16. Stanton, M.L. 2003. Interacting guilds: moving beyond the pairwise perspective on mutualisms. *Am. Nat.* **162:** S10–S23.
17. Thomson, J.D., S.C. Peterson & L.D. Harder. 1987. Response of traplining bumblebees to competition experiments: shifts in feeding location and efficiency. *Oecologia* **71:** 295–300.
18. Gill, F.B. 1988. Trapline foraging by hermit hummingbirds: competition for an undefended, renewable resource. *Ecology* **69:** 1933–1942.
19. Thomson, D. 2004. Competitive interactions between the invasive European honey bee and native bumble bees. *Ecology* **85:** 458–470.
20. Mitchell, R.J., R.J. Flanagan, B.J. Brown, *et al.* 2009. New frontiers in competition for pollination. *Ann. Bot.* **103:** 1403–1413.
21. Caruso, C.M. 2000. Competition for pollination influences selection on floral traits of ipomopsis aggregata. *Evolution* **54:** 1546–1557.
22. Hansen, T.F., W.S. Armbruster & L. Antonsen. 2000. Comparative analysis of character displacement and spatial adaptations as illustrated by the evolution of *Dalechampia* blossoms. *Am. Nat.* **156:** S17–S34.
23. Morales, C.L. & A. Traveset. 2008. Interspecific pollen transfer: magnitude, prevalence and consequences for plant fitness. *Crit. Rev. Plant Sci.* **27:** 221–238.
24. Holland, J.N. 2002. Benefits and costs of mutualism: demographic consequences in a pollinating seed-consumer interaction. *Proc. R. Soc. Lond. Ser. B-Biol. Sci.* **269:** 1405–1412.
25. Morris, W.F., J.L. Bronstein & W.G. Wilson. 2003. Three-way coexistence in obligate mutualist-exploiter interactions: the potential role of competition. *Am. Nat.* **161:** 860–875.
26. Bronstein, J.L. 2001. The exploitation of mutualisms. *Ecol. Lett.* **4:** 277–287.
27. Yu, D.W. 2001. Parasites of mutualisms. *Biol. J. Linn. Soc.* **72:** 529–546.
28. Douglas, A.E. 2008. Conflict, cheats and the persistence of symbioses. *New Phytol.* **177:** 849–858.
29. Raine, N.E., N. Gammans, I.J. Macfadyen, *et al.* 2004. Guards and thieves: antagonistic interactions between two ant species coexisting on the same ant-plant. *Ecol. Entomol.* **29:** 345–352.
30. Internicola, A.I., P.A. Page, G. Bernasconi & L.D.B. Gigord. 2007. Competition for pollinator visitation between deceptive and rewarding artificial inflorescences: an experimental test of the effects of floral colour similarity and spatial mingling. *Funct. Ecol.* **21:** 864–872.
31. Heil, M., M. González-Teuber, L.W. Clement, *et al.* 2009. Divergent investment strategies of *Acacia* myrmecophytes

and the coexistence of mutualists and exploiters. *Proc. Natl. Acad. Sci. USA* **106**: 18091–18096.

32. Irwin, R., J.L. Bronstein, J. Manson & L.E. Richardson. 2010. Nectar-robbing: ecological and evolutionary perspectives. *Annu. Rev. Ecol. Evol. Syst.* **41**: 271–292.

33. Dohzono, I., Y.K. Kunitake, J. Yokoyama & K. Goka. 2008. Alien bumble bee affects native plant reproduction through interactions with native bumble bees. *Ecology* **89**: 3082–3092.

34. Kiers, E.T., T.M. Palmer, A.R. Ives, *et al.* 2010. The global breakdown of mutualistic interactions among species. *Ecol. Lett.* **13**: 1459–1474.

35. Kiers, E.T., R.A. Rousseau, S.A. West & R.F. Denison. 2003. Host sanctions and the legume-rhizobium mutualism. *Nature* **425**: 78–81.

36. Kiers, E.T., R.A. Rousseau & R.F. Denison. 2006. Measured sanctions: legume hosts detect quantitative variation in rhizobium cooperation and punish accordingly. *Evol. Ecol. Res.* **8**: 1077–1086.

37. Kiers, E.T. & R.F. Denison. 2008. Sanctions, cooperation, and the stability of plant-rhizosphere mutualisms. *Annu. Rev. Ecol. Evol. Syst.* **39**: 215–236.

38. Bull, J.J. & W.R. Rice. 1991. Distinguishing mechanisms for the evolution of cooperation. *J. Theor. Biol.* **149**: 63–74.

39. Sachs, J.L., U.G. Mueller, T.P. Wilcox & J.J. Bull. 2004. The evolution of cooperation. *Q. Rev. Biol.* **79**: 135–160.

40. Bshary, R. & J.L. Bronstein. 2011. A general scheme to predict partner control mechanisms in pairwise cooperative interactions between unrelated individuals. *Ethology* **117**: 271–283.

41. Yu, D.W. & N.E. Pierce. 1998. A castration parasite of an ant-plant mutualism. *Proc. R. Soc. Lond. Ser. B* **265**: 375–382.

42. Stanton, M.L., T.M. Palmer & T.P. Young. 2002. Competition-colonization trade-offs in a guild of African Acacia-ants. *Ecol. Monogr.* **72**: 347–363.

43. Crowley, P.H. & J.J. Cox. 2011. Intraguild mutualism. *Trends Ecol. Evol.* **26**: 627–633.

44. Kapan, D.D. 2001. Three-butterfly system provides a field test of Müllerian mimicry. *Nature* **409**: 338–340.

45. Székely, T., T. Szép & T. Juhász. 1989. Mixed species flocking of tits (*Parus* spp.): a field experiment. *Oecologia* **78**: 490–495.

46. Bshary, R. & R. Noë. 1997. Red colobus and Diana monkeys provide mutual protection against herbivores. *Anim. Behav.* **54**: 1461–1474.

47. Bshary, R., A. Kohner, K. Ait-el-Djoudi & H. Fricke. 2006. Interspecific communicative and coordinated hunting between groupers and giant moray eels in the Red Sea. *PLoS. Biol.* **4**: 2393–2398.

48. Ranta, E., H. Rita & K. Lindström. 1993. Competition versus cooperation: success of individuals foraging alone and in groups. *Am. Nat.* **142**: 42–58.

49. Hino, T. 1998. Mutualistic and commensal organization of avian mixed-species foraging flocks in a forest of western Madagascar. *J. Avian Biol.* **29**: 17–24.

50. Yuma, M. 1994. Food-habits and foraging behavior of benthivorous cichlid fishes in Lake Tanganyika. *Environ. Biol. Fish.* **39**: 173–182.

51. Jack, C.N., J.G. Ridgeway, N.J. Mehdiabadi, *et al.* 2008. Segregate or cooperate- a study of the interaction between two species of Dictyostelium. *BMC Evol. Biol.* **8**.

52. Hartnett, D.C., B.A.D. Hetrick, G.W.T. Wilson & D.J. Gibson. 1993. Mycorrhizal influence on intra- and interspecific neighbour interactions among co-occurring prairie grasses. *J. Ecol.* **81**: 787–795.

53. Stachowicz, J.J. & M.E. Hay. 1996. Facultative mutualism between an herbivorous crab and a coralline alga: advantages of eating noxious seaweeds. *Oecologia* **105**: 377–387.

54. Wilson, G.W.T. & D.C. Hartnett. 1997. Effects of mycorrhizae on plant growth and dynamics in experimental tallgrass prairie microcosms. *Am. J. Bot.* **84**: 478–482.

55. Stachowicz, J.J. 2001. Mutualism, facilitation, and the structure of ecological communities. *Bioscience* **51**: 235–246.

56. Callaway, R.M. 2007. *Positive Interactions and Interdependence in Plant Communities.* Springer. Dordrecht.

57. Heath, K.D. & P. Tiffin. 2007. Context dependence in the coevolution of plant and rhizobial mutualists. *Proc. R. Soc. B-Biol. Sci.* **274**: 1905–1912.

58. Sachs, J.L. & E.L. Simms. 2006. Pathways to mutualism breakdown. *Trends Ecol. Evol.* **21**: 585–592.

59. Sachs, J.L. & E.L. Simms. 2008. The origins of uncooperative rhizobia. *Oikos* **117**: 961–966.

60. Hardin, G. 1968. The tragedy of the commons. *Science* **162**: 1243–1248.

61. Noë, R. & P. Hammerstein. 1994. Biological markets – supply-and-demand determine the effect of partner choice in cooperation, mutualism and mating. *Behav. Ecol. Sociobiol.* **35**: 1–11.

62. Bshary, R. & D. Schaffer. 2002. Choosy reef fish select cleaner fish that provide high-quality service. *Anim. Behav.* **63**: 557–564.

63. Simms, E.L. & D.L. Taylor. 2002. Partner choice in nitrogen-fixation mutualisms of legumes and rhizobia. *Integr. Comp. Biol.* **42**: 369–380.

64. Nyholm, S.V. & M.J. McFall-Ngai. 2004. The winnowing: establishing the squid-*Vibrio* symbiosis. *Nat. Rev. Microbiol.* **2**: 632–642.

65. West, S.A., E.T. Kiers, I. Pen & R.F. Denison. 2002. Sanctions and mutualism stability: when should less beneficial mutualists be tolerated? *J. Evol. Biol.* **15**: 830–837.

66. Foster, K.R. & H. Kokko. 2006. Cheating can stabilize cooperation in mutualisms. *Proc. R. Soc. B-Biol. Sci.* **273**: 2233–2239.

67. Bshary, R. & A.S. Grutter. 2002. Experimental evidence that partner choice is a driving force in the payoff distribution among cooperators or mutualists: the cleaner fish case. *Ecol. Lett.* **5**: 130–136.

68. Edwards, D.P. & D.W. Yu. 2007. The roles of sensory traps in the origin, maintenance, and breakdown of mutualism. *Behav. Ecol. Sociobiol.* **61**: 1321–1327.

69. Roberts, G. & T.N. Sherratt. 1998. Development of cooperative relationships through increasing investment. *Nature* **394**: 175–179.

70. Sherratt, T.N. & G. Roberts. 1999. The evolution of quantitatively responsive cooperative trade. *J. Theor. Biol.* **200**: 419–426.

71. Akçay, E. & J. Roughgarden. 2007. Negotiation of mutualism: rhizobia and legumes. *Proc. R. Soc. B-Biol. Sci.* **274:** 25–32.

72. Simms, E.L., D.L. Taylor, J. Povich, *et al.* 2006. An empirical test of partner choice mechanisms in a wild legume-rhizobium interaction. *Proc. R. Soc. B-Biol. Sci.* **273:** 77–81.

73. Kiers, E.T., M. Duhamel, Y. Beesetty, *et al.* 2011. Reciprocal rewards stabilize cooperation in the mycorrhizal symbiosis. *Science* **333:** 880–882.

74. Bshary, R. 2002. Building up relationships in asymmetric co-operation games between the cleaner wrasse *Labroides dimidiatus* and client reef fish. *Behav. Ecol. Sociobiol.* **52:** 365–371.

75. Noë, R. & P. Hammerstein. 1995. Biological markets. *Trends Ecol. Evol.* **10:** 336–339.

76. Akçay, E. & E.L. Simms. 2011. Negotiation, sanctions, and context dependency in the legume-rhizobium mutualism. *Am. Nat.* **178:** 1–14.

77. Axén, A.H. & N.E. Pierce. 1998. Aggregation as a cost-reducing strategy for lycaenid larvae. *Behav. Ecol.* **9:** 109–115.

78. Pellmyr, O. & C.J. Huth. 1994. Evolutionary stability of mutualism between yuccas and yucca moths. *Nature* **372:** 257–260.

79. Bshary, R. & A.S. Grutter. 2002. Asymmetric cheating opportunities and partner control in a cleaner fish mutualism. *Anim. Behav.* **63:** 547–555.

80. Denison, R.F. 2000. Legume sanctions and the evolution of symbiotic cooperation by rhizobia. *Am. Nat.* **156:** 567–576.

81. Friesen, M.L. & A. Mathias. 2010. Mixed infections may promote diversification of mutualistic symbionts: why are there ineffective rhizobia? *J. Evol. Biol.* **23:** 323–334.

82. Johnstone, R.A. & R. Bshary. 2008. Mutualism, market effects and partner control. *J. Evol. Biol.* **21:** 879–888.

83. Holland, J.N., D.L. DeAngelis & J.L. Bronstein. 2002. Population dynamics and mutualism: functional responses of benefits and costs. *Am. Nat.* **159:** 231–244.

84. Holland, J.N. & D.L. DeAngelis. 2006. Interspecific population regulation and the stability of mutualism: fruit abortion and density-dependent mortality of pollinating seed-eating insects. *Oikos* **113:** 563–571.

85. Jaeger, N., F. Pompanon & L. Després. 2001. Variation in predation costs with *Chiastocheta* egg number on *Trollius europaeus*: how many seeds to pay for pollination? *Ecol. Entomol.* **26:** 56–62.

86. Peng, Y.Q., D.R. Yang & Q.Y. Wang. 2005. Quantitative tests of interaction between pollinating and non-pollinating fig wasps on dioecious *Ficus hispida*. *Ecol. Entomol.* **30:** 70–77.

87. Huth, C.J. & O. Pellmyr. 1999. Yucca moth oviposition and pollination behavior is affected by past flower visitors: evidence for a host-marking pheromone. *Oecologia* **119:** 593–599.

88. Horn, K.C. & J.N. Holland. 2010. Discrimination among floral resources by an obligately pollinating seed-eating moth: host-marking signals and pollination and florivory cues. *Evol. Ecol. Res.* **12:** 119–129.

89. Ferdy, J.B., L. Després & B. Godelle. 2002. Evolution of mutualism between globeflowers and their pollinating flies. *J. Theor. Biol.* **217:** 219–234.

90. Jones, E.I. 2009. *Evolutionary Dynamics of Mutualism: The Role of Exploitation and Competition*. PhD thesis, University of Arizona. Tucson, AZ.

91. Wang, R.W., J. Ridley, B.F. Sun, *et al.* 2009. Interference competition and high temperatures reduce the virulence of fig wasps and stabilize a fig-wasp mutualism. *PLoS. One* **4**.

92. Leigh, E.G. 2010. The evolution of mutualism. *J. Evol. Biol.* **23:** 2507–2528.

93. Doebeli, M. & N. Knowlton. 1998. The evolution of interspecific mutualisms. *Proc. Natl. Acad. Sci. USA.* **95:** 8676–8680.

94. Yamamura, N., M. Higashi, N. Behera & J.Y. Wakano. 2004. Evolution of mutualism through spatial effects. *J. Theor. Biol.* **226:** 421–428.

95. Foster, K.R. & T. Wenseleers. 2006. A general model for the evolution of mutualisms. *J. Evol. Biol.* **19:** 1283–1293.

96. Yamamura, N. 1996. Evolution of mutualistic symbiosis: a differential equation model. *Res. Popul. Ecol.* **38:** 211–218.

97. Margulis, L. 1993. *Symbiosis in Cell Evolution*. Freeman. New York.

98. Gargas, A., P.T. Depriest, M. Grube & A. Tehler. 1995. Multiple origins of lichen symbioses in fungi suggested by SSU rDNA phylogeny. *Science* **268:** 1492–1495.

99. Janson, E.M., J.O. Stireman, M.S. Singer & P. Abbot. 2008. Phytophagous insect-microbe mutualisms and adaptive evolutionary diversification. *Evolution* **62:** 997–1012.

100. Sachs, J.L. & J.J. Bull. 2005. Experimental evolution of conflict mediation between genomes. *Proc. Natl. Acad. Sci. USA* **102:** 390–395.

101. Jablonski, D. 2008. Biotic interactions and macroevolution: extensions and mismatches across scales and levels. *Evolution* **62:** 715–739.

102. Muchhala, N., Z. Brown, W.S. Armbruster & M.D. Potts. 2010. Competition drives specialization in pollination systems through costs to male fitness. *Am. Nat.* **176:** 732–743.

103. Tripp, E.A. & P.S. Manos. 2008. Is floral specialization an evolutionary dead-end? Pollination system transitions in *Ruellia* (Acanthaceae). *Evolution* **62:** 1712–1736.

104. Fortuna, M.A. & J. Bascompte. 2006. Habitat loss and the structure of plant-animal mutualistic networks. *Ecol. Lett.* **9:** 278–283.

105. Ashworth, L., R. Aguilar, L. Galetto & M.A. Aizen. 2004. Why do pollination generalist and specialist plant species show similar reproductive susceptibility to habitat fragmentation? *J. Ecol.* **92:** 717–719.

106. Bond, W.J. 1994. Do mutualisms matter – assessing the impact of pollinator and disperser disruption on plant extinction. *Philos. Trans. R. Soc. B-Biol. Sci.* **344:** 83–90.

107. Briand, F. & P. Yodzis. 1982. The phylogenetic distribution of obligate mutualism – evidence of limiting similarity and global instability. *Oikos* **39:** 273–275.

108. Pellmyr, O. & J. Leebens-Mack. 1999. Forty million years of mutualism: evidence for Eocene origin of the yucca-yucca moth association. *Proc. Natl. Acad. Sci. USA* **96:** 9178–9183.

109. Rønsted, N., G.D. Weiblen, J.M. Cook, *et al.* 2005. 60 million years of co-divergence in the fig-wasp symbiosis. *Proc. R. Soc. B-Biol. Sci.* **272:** 2593–2599.

110. Johnson, S.D. & K.E. Steiner. 2000. Generalization versus specialization in plant pollination systems. *Trends Ecol. Evol.* **15:** 140–143.

species and, within each such species, between populations in sympatry and allopatry with the other species.[13] Specifically, most research presumes that divergence is mediated exclusively by genetically canalized changes (i.e., divergence that reflects allelic or genotype frequency changes). This focus on genetic differentiation has seemingly arisen, in part, for two reasons. First, models of character displacement generally assume specific genetic architectures (e.g., single locus vs. multilocus)[8,21] and genetic processes (e.g., mutation, gene flow).[24,26] Empirical tests of the predictions of such alternative genetic models necessarily require investigations of the genetic basis of character displacement (see below for more detail). Second, a focus on genetic differentiation has also emerged because, among the widely accepted criteria used to define character displacement, is the requirement that putative cases of character displacement should reflect genetic differentiation of populations and species.[4,23,27–30]

Although genetically canalized traits have largely been the focus of proximate studies of character displacement, Wilson's[1] quote at the beginning of this paper highlights an alternative mechanism that can underpin displacement: namely, phenotypic plasticity. In the context of character displacement, plasticity is manifest as competitively mediated trait expression (i.e., trait expression that varies depending on the presence of competition). Although not generally considered in the theory of character displacement (except in models involving learning and cultural transmission[31]), plasticity can effectively generate trait divergence between species and thereby serve as an additional axis of variation on which selection can act to promote the evolution of traits that minimize competitive interactions between species. Indeed, because many populations exhibit heritable variation in whether and how individuals respond through environmentally induced change,[32,33] plasticity can itself evolve,[34–36] and plastic traits can therefore satisfy the criteria for demonstrating character displacement.[37,38] Thus, as we describe in greater detail later, evolved environmentally induced niche shifts can constitute character displacement.

Note, however, that genetic canalization and plasticity are not mutually exclusive mechanisms of trait production. Instead, these two proximate mechanisms are best thought of as occupying different positions along a continuum of environmental influences on trait production,[39] with strict genetic control of trait variation at one end of the continuum and with pure environmental induction of trait variation at the opposite end. In reality, most traits have both genetic and environmental components.[40] Moreover, plasticity in one or more traits is often required to hold another trait constant in the face of changing environmental conditions (thereby maintaining homeostasis),[41] further demonstrating the interdependence of these two influences on development.

Perhaps more importantly, these two proximate mechanisms are often evolutionarily interchangeable,[42] meaning that selection can slide trait regulation anywhere along this continuum.[39,41,42] Specifically, when genetic variation for the degree of environmental influence is present, then selection can act on this variation to promote the evolution of either increased or decreased environmental sensitivity.[42] If selection eliminates all environmental influences (i.e., if "genetic assimilation"[43,44] occurs), the end result is a genetically canalized trait (note that selection can also promote "epigenetic assimilation," wherein trait expression becomes less sensitive to environmental influences due to inherited environmental effects,[45] such as a maternal effect,[46] cultural transmission,[47] or parasite transmission[48]).

With genetic assimilation, the evolution of a new trait does not require the emergence of new genes or new gene complexes; instead, selection acts on existing genetic architecture and epigenetic interaction.[32,49,50] In other words, a plastic trait can be converted into a canalized trait (or, alternatively, it can be converted into a trait that shows enhanced plasticity) through evolutionary adjustments in the regulation of trait expression. Experiments have demonstrated such evolutionary shifts in the degree to which populations are sensitive to environmental influences,[34] including the complete loss of plasticity,[44] and numerous examples of evolution by natural selection might be explained by genetic assimilation.[42,51–53]

Genetic assimilation might be a more common mechanism of character displacement than has been heretofore appreciated.[39] We will focus on the potential role of genetic assimilation in character displacement later in this paper. Before we do so, however, we briefly review separately the evidence for genetically canalized traits in character displacement and plasticity, respectively.

Genetic mechanisms of divergence

Competitively mediated selection is expected to target genes that influence the expression of phenotypes involved in resource use or reproduction. Although relatively few studies have thus far identified these selective targets, the studies that have been conducted to date suggest that a diversity of genetic mechanisms can mediate character displacement. Four such mechanisms are briefly reviewed below.

First, character displacement might arise via a "single-allele mechanism."[54–56] With such a mechanism, divergence arises between interacting species because the same allele in both species enhances existing differences between them. An allele that enhances sensory sensitivity to male sexual signals, for example, might render females better at identifying conspecific males and thereby promote reproductive character displacement. For instance, in fruit flies, *Drosophila pseudoobscura* and *D. persimilis*, a single allele potentially mediates species recognition.[57] Females of both species from sympatric populations exhibit greater discrimination against heterospecifics compared to females from allopatric populations.[58] This discrimination ability in both species appears to be at least partly mediated by a single allele at the *Coy-2* chromosomal region that enhances discrimination against heterospecifics (likely by influencing a female's ability to detect species-specific olfactory cues).[57,59] Although additional mechanisms might also be involved in mate discrimination in these flies,[59,60] this system illustrates how costly reproductive interactions between species can be reduced by a single allele shared by both species.

Single-allele mechanisms may mediate ecological character displacement as well. Enhanced sensory sensitivity might also refine preexisting food preferences within different species and thereby reduce overlap between them in resource use. Moreover, a single allele might reduce dispersal tendencies, which could exaggerate differences between species in habitat use and thereby preclude interactions between them. Such possibilities will remain speculative, however, until more is known regarding the proximate mechanisms underlying traits that are involved in minimizing resource competition between species.

A second, slightly more complex mechanism—but one still involving a single locus—entails alternative alleles at a locus between the interacting species that specifies divergent resource-use or reproductive traits. Such allelic differentiation at a single locus might suffice to reduce competitive or reproductive interactions between species. For example, passion-vine butterflies (genus *Heliconius*) show great diversity in wing color patterns. This genus has undergone rapid speciation,[61] but many of its constituent species have also converged in wing color pattern owing to Müllerian mimicry.[62,63] However, for species that use wing coloration in mate choice (such as *H. cydno* and *H. pachinus*, where males use wing color to discriminate conspecific from heterospecific mates[64]), such convergence in wing color pattern increases the risk of hybridization.[65] Thus, interacting species are expected to undergo reproductive character displacement as a means of reducing such costly interactions. Recent research suggests that a single locus might encode both mate preference and wing coloration.[64] Because pigments involved in wing coloration also occur in the eye (and affect perception of wing coloration),[64] a single gene affects both coloration and perception of—and preference for—that coloration.[66] Therefore, species that possess alternative alleles at a single locus can diverge in traits such as sexual signaling and mate choice that mediate character displacement. If that locus has pleiotropic effects, as in *Heliconius* butterflies, then divergence in multiple traits can arise simultaneously.

In contrast to mechanisms (such as those above) that depend either on a single allele or on alternative alleles at a single locus (with or without pleiotropy), a third mechanism arises when character displacement is under the control of multiple, divergent loci. A possible example comes from pied flycatchers (*Ficedula hypoleuca*) and collared flycatchers (*F. albicollis*), which have undergone reinforcement of female preferences and male coloration.[67] Such divergence appears to be mediated by multiple, divergent loci.[68] Similarly, character displacement between benthic and limnetic species of stickleback fish (*Gasterosteus aculeatus* complex[27]) involves multiple traits that appear to be encoded by numerous loci: F_1 hybrids are intermediate in morphology between the two parent species, and these differences persist over multiple generations in a common laboratory environment.[69–72]

At present, it is unclear how commonly character displacement arises through the different mechanisms above. Theoretical models suggest that character displacement (reproductive character displacement or reinforcement, in particular) will be more likely to arise under a one-allele mechanism than a two-allele mechanism.[55] However, other models suggest that character displacement can also arise when mediated by multiple loci.[21,22]

Given that character displacement typically entails complex suites of traits,[73,74] one might expect that competitively mediated selection would target multiple loci. Nevertheless, a key (theoretical and empirical) challenge has been to explain how such a mechanism would persist in the face of recombination, which would tend to break up coadapted gene complexes that encode species differences.[75,76] This is particularly problematic in species that exchange genes (as has been documented in species that have undergone character displacement, including sticklebacks,[77] spadefoot toads,[78] Darwin's finches,[79] and nightingales[80]), because hybridization tends to "scramble" allelic combinations at loci that isolate species. One possibility is that the loci encoding species differences might reside in areas that are protected from recombination,[75,76,81,82] such as inside chromosomal inversions,[81] near the chromosome's centromere,[82] or on sex chromosomes.[67]

In the three previous mechanisms (i.e., single-allele, multiple alleles at a single-locus, and multiple-loci mechanisms), the genes involved in mediating character displacement encode for proteins that produce the trait(s) that undergo divergence (e.g., cuticular hydrocarbons used in mate recognition in fruit flies; color pigments in butterflies). In other words, character displacement involves changes in the protein-encoding regions of the genome. Yet, differences in protein-coding sequences are not the only means by which species could become differentiated.[83–88] Instead, species might diverge in cis-regulatory (noncoding) regions that are involved in resource use or reproduction. Thus, a fourth mechanism involves divergence in regions of the genome that regulate the expression of the genes that encode for displaced traits between species.

A putative example comes from Darwin's finches, which have undergone both ecological and reproductive character displacement in the size and shape of their beaks[89,90] (the size and shape of an individual's beak determines not only which resources it can use,[89] it can also affect the production of male song, which is used in territorial defense and mate attraction).[91]

Recent studies of how bird beaks form developmentally point to the possible genetic targets of divergent selection in this system. Beak development is influenced by several genes that encode a series of signaling molecules, including fibroblast growth factor 8 (*Fgf8*) and sonic hedgehog (*Shh*). These two gene products influence the expression of a third signaling molecule, bone morphogenetic protein 4 (*Bmp4*), which governs differences between species in beak depth.[92,93] These two gene products also affect the expression of the gene calmodulin (*CaM*), which encodes for a calcium-binding protein involved in apoptosis that influences beak length.[93]

Although any one of these four genes could conceivably serve as a target of selection during character displacement, these genes are also involved in several crucial metabolic processes (e.g., *CaM* is involved in numerous, vital cellular processes). Thus, any changes in these genes would likely have deleterious consequences. Consequently, it is unlikely that selection would favor alternative alleles at these genes in different species and populations. Instead, character displacement has likely occurred when selection brought about changes in the regulation of these genes.[89]

Regulatory mutations are similarly thought to be involved in competitively induced divergence in pheromones (specifically, cuticular hydrocarbons) that are used in mate choice in the *Drosophila serrata* complex.[94] Such regulatory mutations have also been implicated in mediating reproductive character displacement (specifically, floral-color divergence) in the Texas wildflower, *Phlox drummondii*.[95] Although the role of regulatory mutations during adaptation remains controversial,[96,97] such mutations are increasingly viewed as being important during adaptive population divergence.[88,98]

In sum, studies of the genetic targets of competitively mediated selection have revealed a diversity of genetic mechanisms, which appear to differ in the degree to which they facilitate character displacement. However, we still do not know whether one mechanism is more prevalent than the others in mediating character displacement. Although putative examples exist for each mechanism, additional work across a diversity of systems is critically needed to determine whether some genetic mechanisms are more likely to underpin character displacement than others as predicted by theory.[55] In the next section, we consider an alternative mechanism that might be effective at mediating character displacement: phenotypic plasticity.

Mechanisms of environmentally induced divergence

In contrast to reflecting genetically canalized differences, divergent traits might alternatively arise through phenotypic plasticity. Specifically, rather than being produced constitutively (as with the genetic mechanisms described earlier), divergent traits might be expressed facultatively, such that they are produced in an individual only when it experiences competition from a heterospecific (e.g., see Fig. 1). Although relatively few studies have explicitly considered plasticity's role in character displacement (notwithstanding Wilson's[1] quote at the outset of this paper), such environmentally contingent niche shifts may play an underappreciated role in mediating character displacement.[13,14,99] Here we discuss two different mechanisms by which plasticity may promote character displacement. Before we do so, we begin with a caveat.

In particular, competitively mediated plasticity may or may not constitute character displacement.[37,38] Character displacement is defined as trait evolution that arises as an adaptive response to resource competition or deleterious reproductive interactions between species.[4,5,9,13,14,100] Competitively mediated plasticity therefore constitutes character displacement only when it has actually evolved in direct response to competitively mediated selection and when it lessens competitive interactions between heterospecifics.

Even when environmentally induced traits have not themselves undergone character displacement, they can still be critical to the process. Specifically, lineages that respond to heterospecific competitors through facultative adjustments in their phenotype may be less likely to go extinct via competitive or reproductive exclusion. Indeed, plasticity is increasingly viewed playing a critical role in shielding populations from extinction when confronted with changing environments.[39,101,102] In the context of competitive interactions, plasticity might enable individuals to produce resource-use or reproduc-

tive phenotypes that are less like the phenotypes expressed by their competitors. In this way, competitively mediated plasticity might provide a mechanism for reducing the frequency and intensity of competitive interactions. Thus, even in the absence of character displacement, plasticity may promote species coexistence through resource partitioning[103] or reproductive partitioning. Such a process could serve to promote developmentally mediated species sorting, in which species express different phenotypes at contact, but not owing to preexisting canalized differences.[104,105]

Many organisms appear to possess the ability to respond to heterospecifics through facultative adjustments of their phenotype.[42,103,106–109] For example, when confronted with a heterospecific competitor, the individual members of many species of plants facultatively express traits that reduce interspecific competition, such as altering the spatial positioning of their roots,[110–113] modifying the width of their leaves,[114] or adjusting their physiology.[115] Similarly, when faced with a heterospecific competitor, many species of fish[109] and amphibians[37,38] facultatively express resource-use morphologies that differ from those expressed by the heterospecific. Some species can even facultatively adjust their mate preferences in the presence of heterospecifics, thereby reducing costly reproductive interactions between species.[116] Of such cases, several systems have illustrated that plasticity does mediate character displacement,[37,38,99,109,117,118] via two different mechanisms.

First, plasticity may underlie character displacement when competitively mediated selection leads to the evolution of a reaction norm that minimizes competitive interactions. Under this mechanism, selection acts on underlying heritable variation in either the tendency to respond to competitors in the first place or the manner in which individuals express these responses (or both) in propelling competitively mediated trait evolution (i.e., character displacement). Such evolved shifts in environmentally induced resource-use or reproductive traits can satisfy the widely accepted criteria for demonstrating character displacement.[37] In particular, experiments have demonstrated that these induced shifts are indeed caused by the presence of a heterospecific competitor *per se* (Fig. 1) and that they lessen costly interactions with the heterospecific that induces them.[37,116] Furthermore,

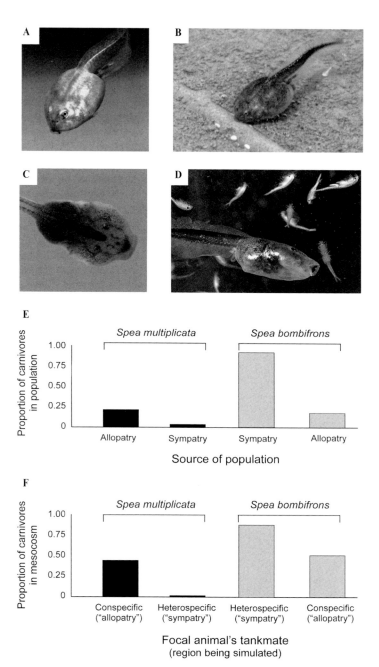

Figure 1. An experimental demonstration of how phenotypic plasticity might mediate ecological character displacement. Spade-foot toad tadpoles (*Spea multiplicata* and *S. bombifrons*) typically occur as (A) an omnivore morph, (B) which specializes on plant material and detritus, and (C) a morphologically distinct carnivore morph, (D) which specializes on, and is induced by, anostra-can fairy shrimp. (E) In allopatric populations, each species produces similar, intermediate frequencies of both morphs. However, in sympatry, *S. multiplicata* shift to producing mostly omnivores, whereas *S. bombifrons* shift to producing mostly carnivores; that is, these two species have undergone character displacement in morph production. (F) Similar niche shifts can be experimentally recreated in the lab. When allopatric individuals are reared with a single conspecific and fed limited amounts of both shrimp and detritus, they produced similar proportions of both morphs (as in natural allopatric populations; see panel E). By contrast, when the two species are experimentally combined, *S. multiplicata* shift to producing mostly omnivores, whereas *S. bombifrons* shift to producing mostly carnivores (as in natural sympatric populations). These facultative niche shifts appear to reflect differences between species in ability to capture and consume shrimp, an environmental cue that induces carnivores.

experimentally induced shifts often mirror, in magnitude and direction, the fixed phenotypic differences between naturally occurring sympatric and allopatric populations (e.g., see Fig. 1). Finally, facultative shifts have been shown to evolve in sympatry versus allopatry.[38,116]

A second general mechanism by which plasticity may promote character displacement in when it is transmitted reliably across generations and thereby forms the basis of an alternative inheritance system on which adaptive evolution can unfold. There are two types of transgenerational plasticity that may be important in mediating character displacement. The first are maternal effects, which occur when a female's phenotype influences her offspring's phenotype, independent of the direct effects of her coding sequences on her offspring's phenotype.[46,119–123] Because they can be induced by interspecific competition,[117,120] mediate adaptive phenotypic change,[46,121,122,124,125] and be transmitted reliably across generations,[126] maternal effects might play an underappreciated role in promoting character displacement.[117] A second form of transgenerational plasticity is cultural inheritance, mediated by learning. In many animals, for example, preference for conspecific mates is learned.[89,127,128] Likewise, many animal species learn to use new food resources.[129,130] Once a population acquires such learned mate or food preferences, these preferences can be transmitted across generations and even reinforce differences between species, thereby mediating ecological or reproductive character displacement.[127,130] Such a situation is illustrated in Darwin's finches, where learning played a crucial role in mediating reproductive character displacement.[127]

In sum, plasticity may play a role in mediating character displacement. However, more empirical and theoretical work is needed to determine whether and how each of the above mechanisms contributes to character displacement. For the remainder of the paper, we consider some ways in which plasticity may play an important role in character displacement. In particular, we contrast plasticity and genetic canalization in terms of their effects on the rapidity with which character displacement can occur. Although the two mechanisms are by no means mutually exclusive, they differ in important ways that can ultimately affect when and how character displacement occurs.

Tempo and mode of character displacement

As we highlighted in the Introduction, character displacement is a form of adaptive evolution, and the earlier mechanisms are therefore not unique to character displacement. Nevertheless, the mechanisms mediating character displacement may determine when and how it occurs. In this section, we focus on this issue.

We first consider each mechanism's impacts on the speed of character displacement. We do so, because character displacement is a time-limited process. Populations experiencing competition are often at risk of competitive or reproductive exclusion. Therefore, populations that do not respond sufficiently rapidly to the presence of competition risk extinction. Indeed, a meta-analysis by Schluter[5] indicated that some taxa are more likely to undergo character displacement than others. Although such variation may reflect researcher bias in terms of the systems chosen to study character displacement, such a pattern may also indicate that character displacement is more likely to occur in some conditions versus others. One such condition that may be important is the ease and rapidity with which character displacement takes place. Thus, consideration should be given to the differential rate by which genetically canalized versus plastic traits are likely to evolve, in addition to the other conditions that appear to facilitate character displacement (e.g., strong selection, certain genetic architectures, and initial trait differences).[21,25,131,132] Such differences in the speed of evolution for these different proximate mechanisms could therefore explain when character displacement is more—or less—likely to occur.

Why proximate mechanisms likely differ in speed of divergence

A key factor that determines the speed of character displacement is the amount of standing variation in the trait(s) under divergent selection.[21,131,132] When abundant standing variation is present in a population that encounters a heterospecific competitor, competitively mediated selection could act on this variation, filtering out those variants that are the most similar to the heterospecific competitor and preserving those that are most dissimilar. This process fuels character displacement,[13,132] and, if standing variation is abundant and selection is strong, then the rate of such an evolutionary response could be rapid in that it could transpire in a handful of generations (Darwin's finches provide a possible example of such rapid evolution[133]).

When standing variation is depleted (or is initially absent), however, character displacement, like any other form of adaptive evolution, may be precluded.[5,132,134] In particular, for character displacement to proceed, new variants must be introduced into the population. If divergence depends entirely on genetically canalized differences, new variants must be introduced into the population via mutation, recombination, or gene flow from another conspecific population or another species (through hybridization).

All three of the above means of acquiring new variants are likely to either transpire relatively slowly or, ironically, simultaneously counteract character displacement even as they provide the raw material necessary for character displacement to proceed. For example, the waiting time for favorable, new mutations to arise and spread in a population can take many generations.[135] Indeed, many genetic models of character displacement predict that the pace of evolution will be slow, especially during the early phase of divergence.[21,22] Yet, if character displacement transpires slowly, the risk of competitive or reproductive exclusion increases.[13] Moreover, although recombination and gene flow can potentially introduce favorable, new genetic variants into a population more quickly than does mutation, these processes will typically affect only a few individuals in any given generation. Perhaps more critically, theory has shown that high rates of gene flow and recombination actually reduce a population's ability to respond to competitors, because maladaptive gene combinations over time swamp adaptive combinations.[26] Thus, even though variants might be introduced that are favored by competitively mediated selection, the population may continue to receive an influx of gene combinations that ultimately prevent local adaptive to competition.

By contrast, divergence driven by phenotypic plasticity might promote a rapid adaptive response to competitors, for at least three reasons. First, an important consequence of environmentally induced phenotypes is that their expression can reveal cryptic genetic variation—genetic variation that is normally not visible as phenotypic variation.[136] Indeed, hidden reaction norms (variation that is pheno-

typically expressed only after an organism experiences changes in environmental conditions) may store an evolutionarily significant pool of cryptic genetic variation upon which selection may act.[33] Therefore, with plasticity, abundant standing variation (which could fuel divergent trait evolution) can be released and exposed to selection as soon as a population experiences competition.[33,137] Essentially, the waiting time between when a population begins to experience competitively mediated selection and when it begins to express variation on which selection can act to produce an adaptive response to competitors is negligible—the process by which new variants are produced occurs over developmental time or over the course of an individual's lifetime (as with learning) rather than over generations.

Second, environmentally induced changes typically occur in numerous individuals simultaneously[42] (especially when a population experiences exploitative competition). This situation contrasts markedly with a mechanism in which the production of new divergent traits requires the introduction of new genetic variants, which typically arise in just a few, or even just one, individual (see above). Moreover, because individuals often differ in whether and how they respond to an environmental cue (such as the presence of a heterospecific competitor), this variation increases the likelihood and speed of an evolutionary response. Although this line of argument has been developed most prominently by West-Eberhard[42,108,138,139] to explain rapid adaptive evolution generally, the application of these ideas to the realm of character displacement could be fruitful, because it potentially broadens the conditions under which character displacement is predicted to occur.

A third reason why plasticity might mediate rapid character displacement is that the ability to respond to competition through plasticity may already exist in a given population, having evolved as an adaptive response to intraspecific competition. For example, forms of plasticity present within species that may mediate character displacement between species include phytochrome-mediated shade-avoidance responses in plants,[114] resource polymorphism in many species of fish and amphibians,[140] and mating polymorphism in many invertebrate and vertebrate species.[141,142] Note that, in some populations, intraspecific competition may also favor genetically canalized variation, such as a genetic polymorphism,[143,144] on which competitively mediated selection can act. However, unless this constitutively expressed variation is maintained in a population by selection, it is at risk of being lost over time. By contrast, with environmental induction, unexpressed (cryptic) genetic variation can be "protected" from selection or chance loss until it is released by a change in the environment.[33]

The presence of preexisting plasticity (and underlying cryptic genetic variation) is important, because as noted in the Introduction, the evolution of a new divergent trait (such as an alternative morphology or behavior that reduces competitive interactions with a heterospecific) need not require new genes or new gene complexes. Instead, selection can repurpose existing genetic pathways and developmental mechanisms and thereby drive rapid, divergent trait evolution. In the context of character displacement, if a population has already evolved the ability to respond to intraspecific competition via environmentally induced changes, the same mechanisms involved in such a response might be co-opted to mediate adaptive divergence between heterospecific competitors (i.e., character displacement). Indeed, empirical data exist (described later) to suggest that intraspecific variation generated by plasticity might form the basis for interspecific variation during character displacement. Essentially, preexisting plasticity (and the underlying developmental mechanisms) might enable character displacement to evolve rapidly along lines of least resistance.[145]

Above, we contrasted the speed with which character displacement arises via genetic canalization versus competitively mediated plasticity. However, as noted earlier, these alternative proximate mechanisms of character displacement are not mutually exclusive and will potentially evolve in concert. Yet, because of the difference in speed with which they can occur, plasticity may initially underpin competitively mediated divergence, with genetic canalization evolving subsequently.[1,118] We describe how this sequence of events might unfold in the next section.

Evolution of character displacement: plasticity first?

Evolutionary biologists have long hypothesized that phenotypic plasticity might precede, and even promote, the genetically canalized traits that arise during adaptive evolution.[32,39,42,44,53,138,146–152] This "plasticity-first" hypothesis rests on the argument that, if selection acts on quantitative genetic variation regulating the expression of initially environmentally dependent traits, it can lead to the evolution of either enhanced plasticity or, alternatively, even the loss of plasticity, that is, genetic assimilation.[44] Such loss might result as an incidental by-product of selection favoring a single phenotypic trait (or an extreme phenotype in a continuum) to the exclusion of all other phenotypes.[43,44,153] Alternatively, plasticity might be lost when the alleles that underlie it are lost through stochastic processes. In particular, when an environmental change results in one trait rarely being produced in a population (such as when, in the presence of a competitor, an individual fails to produce a resource-use or reproductive trait expressed in the competitor's absence; Fig. 1), then alleles that regulate expression of this "hidden" phenotype would not be exposed to selection and would therefore be at greater risk of chance loss.[154,155]

That such genetic assimilation could serve as a mechanism of character displacement has not been heretofore generally appreciated,[13,14] despite Wilson's[1] suggestion, quoted at the outset of the paper. Yet, consider two species that possess preexisting plasticity in resource-use or reproductive phenotypes. Once in sympatry, each species might facultatively express a different subset of the initial phenotypes (Fig. 2). Over time, each species might lose plasticity and become fixed (either through selection or chance) for a different alternative phenotype.[152] Character displacement might begin with an initial phase in which each species evolves the expression of competitively induced plastic phenotypes and then subsequently transition to genetically canalized traits that minimize competitive interactions with the alternate species (Fig. 2).

This plasticity-first hypothesis, of course, is not the only way in which character displacement might evolve—plasticity might not play any role in some species or populations, whereas in other species, plasticity might play a solitary role by mitigating against selection for any further genetically canalized divergence.[156] However, character displacement might be first mediated by plasticity for two reasons. First, those populations that initially express plasticity might be more likely buffered from extinction while genetically canalized

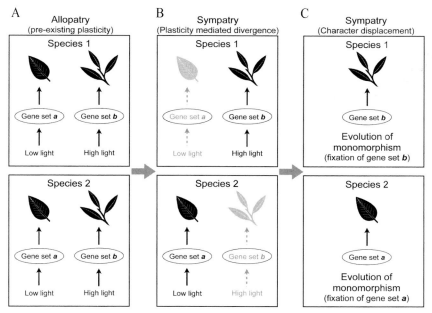

Figure 2. How character displacement might evolve from an initial phase in which trait divergence is environmentally induced to one in which divergence is based on genetically canalized differences and thereby becomes expressed constitutively. (A) Initially, two interacting species may express plasticity in resource-use or reproductive phenotypes. In this case, two species of plants facultatively produce different-sized leaves as an adaptive response to different light levels. (B) When they come into sympatry and compete (e.g., for photons), each species may begin to facultatively express a different subset of the initial phenotypes. Here, Species 1 overtops Species 2, thereby gaining increased access to photons, which triggers the facultative production in Species 1 of small leaves only. By contrast, because it is shaded by Species 1, Species 2 facultatively produces larger leaves only (in each case, the expressed phenotype—and underlying genes—are shown in black, whereas the unexpressed one is shown in gray). If individuals within the same population harbor genetic variation in the degree to which they respond to environmental cues, and if these reaction norms evolve in response to competitively mediated selection and thereby minimize competition between species, then such evolved environmentally induced shifts would constitute character displacement. (C) Over time, both species may lose this preexisting plasticity and become fixed for a different alternative phenotype, possibly because of the loss (through selection or chance) of alleles or gene combinations underlying the nonexpressed phenotype. Thus, character displacement might proceed through an initial phase in which trait divergence is environmentally induced to a later phase in which divergence becomes genetically canalized. Essentially, during character displacement, each species may evolve from expressing a wide range of phenotypes to becoming genetically canalized for a narrower range of phenotypes (in this case, as each species evolves from being polymorphic for leaf shape to being monomorphic).

traits evolve that harden the distinction between species.[13,14] When faced with a new, superior competitor, populations lacking the ability to respond through plasticity might simply undergo reproductive or competitive exclusion before canalized differences evolve.[157] Second, even in populations where plasticity successfully minimizes competitive interactions with heterospecifics, genetic canalization of such traits might be favored if plasticity has costs.[158,159]

This plasticity-first hypothesis has two key implications for the way in which character displacement occurs. First, character displacement may not always proceed slowly. As noted previously, although most

models of character displacement predict that divergence will occur slowly, observations of natural populations have revealed that character displacement can proceed rapidly (i.e., within a few generations).[78,133,160–164] This discrepancy between the models and the data may arise, because most models assume that divergence is based on genetically canalized differences, which (as noted earlier) can evolve slowly. If, instead, divergence is mediated by plasticity, then character displacement might generally proceed rapidly. Although such rapid divergence can also be explained in terms of abundant standing genetic variation (see above), the incorporation of plasticity into models of character displacement could increase the parameter space over which character displacement can occur.

Second, character displacement that is mediated by plasticity and followed by genetic assimilation should generate repeated evolution of the same ecotype (specifically, one that minimizes interspecific competition) in multiple, independently evolving sympatric populations. Such parallel evolution of reproductive or ecological traits among independently evolving sympatric populations that are experiencing similar selective pressures from competition ("parallel character displacement") has indeed been documented in a number of systems.[165–170] Although the plasticity-first hypothesis does not uniquely predict parallel character displacement (standing variation in genetically canalized traits could respond similarly to the same selective pressures across different populations),[166] if stochastic processes (mutation, recombination) serve as the basis of variation in genetically canalized traits that mediate character displacement, then such parallel evolution becomes less likely: populations will be more likely to evolve different routes in response to competitors. By contrast, with preexisting plasticity, the same sets of phenotypes are repeatedly revealed when individuals in different, independent populations experience similar selective pressures from competition. Thus, the environment might play a critical role in not only exerting parallel selection pressures in different populations, but also in generating parallel distributions of traits on which selection acts.[14,39]

Empirical support for the plasticity-first hypothesis

Several systems in which character displacement has been documented appear to conform to the plasticity-first scenario. Specifically, we highlight research on three systems, in which competitively induced plasticity appears to have preceded the evolution of canalized genetic differences.

The first case comes from stickleback fish (*G. aculeatus* complex). Two species occur together in certain small coastal lakes in southwestern Canada: one expresses a "benthic" phenotype, whereas the other expresses a "limnetic" phenotype.[171] These two species are thought to have arisen following the invasion from the ocean of an ancestral limnetic ecotype into lakes that already contained an intermediate ecotype. Following this invasion, both ecological and reproductive character displacement resulted in the parallel evolution of a new benthic ecotype within each such lake, which replaced the ancestral, intermediate ecotype.[171]

Recall from above that the phenotypic differences between these benthic and limnetic ecotypes appear to reflect genetically canalized differences. Recent experiments have revealed, however, that diet-induced plasticity is present in marine sticklebacks,[172] which are thought to represent the ancestral colonists.[171] More importantly, this diet-induced plasticity generates phenotypes that resemble the benthic and limnetic ecotypes found in modern-day (derived) freshwater populations.[172] This recent work indicates that trait divergence between ecotypes was initially environmentally induced, but ultimately may have become genetically canalized. Such plasticity in resource-use traits has been similarly detected in dozens of species of freshwater fish,[109,173–175] including in many that have undergone character displacement,[109] indicating that this route to character displacement is likely not unique to sticklebacks.

A second such example comes from two species of North American spadefoot toads (*Spea multiplicata*, *S. bombifrons*), which have undergone ecological character displacement in tadpole trophic morphology.[37,38,99,163,169,176] Where they co-occur, almost all individuals of one species (*S. multiplicata*) develop into an omnivore morph, which feeds mostly on detritus, whereas almost all individuals of the other species (*S. bombifrons*) develop into a distinctive carnivore morph, which specializes on eating anostracan shrimp and other tadpoles (Fig. 1).

The proximate mechanism of character displacement appears to differ for these two species. In *S. multiplicata*, the phenotypic differences between sympatric and nearby allopatric populations appear to reflect a condition-dependent maternal effect.[117] By contrast, in *S. bombifrons*, the phenotypic differences between sympatric and allopatric populations appear to reflect genetically canalized differences.[118]

In allopatric populations (i.e., ancestral, predisplacement populations), however, both species have the capacity to respond adaptively to the presence of the other species through phenotypic plasticity. Yet, despite the fact that individuals from allopatric populations of both species have the capacity to produce both carnivores and omnivores,[99,163] when allopatric *S. multiplicata* are experimentally exposed to *S. bombifrons*, they facultatively produce mostly omnivores (Fig. 1F), which is similar to the pattern of morph expression found among *S. multiplicata* in naturally occurring, populations in sympatry (Fig. 1E). Conversely, when allopatric *S. bombifrons* are experimentally exposed to *S. multiplicata*, they facultatively produce mostly carnivores (Fig. 1F), which is similar to the pattern of morph expression found among *S. bombifrons* in naturally occurring, populations in sympatry (Fig. 1E). Finally, phylogenetic analyses have found that sympatry is the derived condition in this system.[169] Thus, experimentally initiated niche shifts in ancestral (allopatric) populations mirror in direction and magnitude the canalized shifts observed in derived (sympatric) populations.

Because plasticity in allopatry is ancestral, these data suggest that character displacement in spadefoots might have evolved from an initial phase in which trait divergence was environmentally induced (as observed in modern-day allopatric populations) to one in which divergence became either developmentally canalized (as in modern-day sympatric populations of *S. multiplicata*, where character displacement is mediated by a maternal effect[117]) or genetically canalized (as in modern-day sympatric populations of *S. bombifrons*, where character displacement appears to be mediated by genetic shifts[118]). Presumably, sympatric populations of these two species differ in the mechanism of character displacement because they differ in the length of time that each has been in sympatry,[118] with sympatric populations of *S. bombifrons* having been in sympatry longer than sympatric populations of *S. multiplicata* (due to a range expansion by *S. bombifrons* into the region where character displacement occurs).[177]

Spadefoots therefore highlight two points: (1) they are consistent with the plasticity-first hypothesis for character displacement; and (2) they show that canalized differences need not be strictly genetic. Regarding the second point, a maternal effect in *S. multiplicata* (a form of transgenerational plasticity[178]) has led to canalized differences between the species in morph production in sympatry. Thus, rather than being genetically canalized, the phe-

notypic differences in *S. multiplicata* appear to be epigenetically canalized. As noted earlier, epigenetic assimilation (such as that mediated by a maternal effect) may be an evolutionary precursor to genetic assimilation (recall from above that sympatric populations of *S. multiplicata* have only recently experienced competitively mediated selection imposed by *S. bombifrons*).

Finally, a third example consistent with the plasticity-first hypothesis for character displacement comes from *Anolis* lizards from the Greater Antilles. When different species occur together, they differ in microhabitat use.[179,180] Specifically, on islands where they are sympatric, different species of *Anolis* occur as different ecomorphs, which partition their habitat in different ways, residing (for example) in tree crowns (crown-giant ecomorph), on trunk/crowns (trunk-crown ecomorph), on the trunk/ground (trunk-ground ecomorph), and on twigs (twig ecomorph).[179,180] Different ecomorphs tend to differ in body size and limb length, and manipulative experiments have revealed that at least two species possess the ability to facultatively change limb length in response to different-diameter perches.[181]

Although species differences in limb length might have evolved (at least partly) through a process of species sorting[105] (in which species differences might have arisen through the differential invasion into a habitat of species that happen to differ, or through the differential extinction of species that happen to be too similar to coexist), some differences also have potentially evolved through character displacement,[105,182] and this process might have begun with environmentally induced niche shifts.[181] This system therefore serves as one in which the plasticity-first hypothesis could be tested more extensively.

In sum, several case studies provide evidence that is consistent with the plasticity-first hypothesis for character displacement. Nevertheless, much more work is needed to assess how general this mechanism is for explaining character displacement. As we discuss in more detail later, additional theoretical and empirical work is needed to fully consider plasticity's role in character displacement, and to determine if competitively mediated plasticity ultimately leads to the sort of genetically canalized traits described in the section earlier on genetic mechanisms of divergence.

Conclusions and future directions

Despite character displacement's importance in helping to explain how new species arise, diversify, and coexist,[5,6,13,14] relatively little is known of the proximate mechanisms that mediate character displacement. Indeed, from a proximate perspective, the focus in studies of character displacement has been on understanding how—and what kinds of—genetically canalized differences evolve between populations and species.[5] Yet, increasing evidence suggests that character displacement can alternatively arise through competitively induced phenotypic plasticity.[13,14] It is important to recognize that plasticity can mediate character displacement, because, in some populations and species, plasticity might represent the sole means by which adaptive, divergent traits arise. Perhaps more importantly, from an evolutionary perspective, character displacement might often unfold rapidly as it transitions from an initial phase in which species differences arises through phenotypic plasticity to one in which such divergence is underlain by genetically canalized differences.[118]

With recent theoretical and technical advances, now is a propitious time to both identify the mechanisms of character displacement and critically evaluate how different mechanisms interact during character displacement. Indeed, a number of key questions regarding the proximate basis of character displacement remain unanswered. Here, we highlight five specific questions that promise to provide fruitful avenues for future research.

First, what are the proximate mechanisms of character displacement in diverse taxa? At present, much of what we know comes from studies of relatively few organisms. Additional research on a greater variety of taxa is needed to clarify whether the mechanisms that have already been identified are present in diverse taxa, or whether these mechanisms are unique to the few taxa that have been studied thus far. Generally, there is insufficient data upon which to conclude whether some mechanisms are more prevalent than others.

Second, do different proximate mechanisms generate differences in the ease with which (and, hence, the likelihood that) character displacement occurs? In contrast to genetically canalized differences, shifts underlain by phenotypic plasticity might generally occur more rapidly and affect many more individuals simultaneously.[42] Consequently, because competitively mediated plasticity buffers populations from extinction via competitive or reproductive exclusion,[157] it might increase the likelihood that character displacement will transpire to completion.[13] Additional theoretical and empirical approaches are required to evaluate these predictions. For example, experimental or comparative studies could be used to determine whether, when confronted with a novel competitor, populations (or species) consisting of more plastic genotypes undergo character displacement more rapidly (and more readily) than those consisting of less plastic genotypes. Indeed, this question would be ideally suited for experimental evolution studies with rapidly evolving organisms, such as microbes.[183,184]

Third, does phenotypic plasticity sometimes impede character displacement? Evolutionary biologists have long argued that phenotypic plasticity can dampen selection for diversification, for two reasons.[156] First, plasticity allows a single genotype to produce multiple phenotypes in response to different environmental (and hence, selective) regimes, and thus genetic alternatives are not required for attaining alternative fitness optima, such as those associated with populations in sympatry with a competitor versus those in allopatry. Second, different genotypes can produce the same phenotype via plastic responses, thereby hiding genetic differences between them from selection. Future theoretical and empirical research should seek to evaluate whether there are conditions under which competitively induced plasticity impedes, rather than facilitates, character displacement. In other words, an organism might be so plastic in its responses to competitors that no further evolution transpires. In such a case, plasticity may be more likely to mediate species sorting rather than character displacement.

Fourth, can competitively induced phenotypes lose their environmental sensitivity over evolutionary time and thereby undergo genetic assimilation? Experimental studies with rapidly evolving organisms might also prove informative for addressing this question.[34] Moreover, additional work comparing ancestral populations with derived populations that have undergone character displacement would be valuable in addressing this question.

Finally, does character displacement proceed initially by the evolution of plastic traits and subsequently transition to a phase in which genetically canalized traits evolve and mediate competitive interactions between species? If so, systems in which contact is recent should be more likely to display character displacement in plastic traits, whereas in systems where contact is older, trait differences should be mediated by genetically canalized differences. Addressing this issue is contingent, in part, on answering the fourth question above. Nevertheless, future research should aim to identify additional systems in which to evaluate the plasticity-first hypothesis for the evolution of character displacement. Only by doing so it will be possible to ultimately determine the conditions under which this route occurs and how common it is.

In addressing the above questions, equal emphasis should be placed on doing so in the contexts of both ecological character displacement and reproductive character displacement. Whether the relative importance of alternative mechanisms, and the nature of their interaction, is the same for these two forms of character displacement is itself an open question.

Additional research into the proximate bases of character displacement promises to have far-reaching ramifications. Indeed, because character displacement is central in the origins, abundance, and distribution of biodiversity,[5,6,13,14] understanding its causes can help illuminate some of the most fundamental issues in evolutionary biology and ecology, including how new species arise, how they diversify, and how they coexist.

Acknowledgments

We thank Mary Jane West-Eberhard, Chris Willett, Matt Wund, the members of the Pfennig lab, and two anonymous referees for helpful comments on earlier drafts of the manuscript. We also thank Tim Mousseau and Chuck Fox for inviting us to contribute this paper. Our research on character displacement is funded by grants from the National Science Foundation to D.W.P. and a National Institutes of Health Office of the Director New Innovator Award to K.S.P. A modified version of this paper will appear in Pfennig and Pfennig.[185]

Conflicts of interest

The authors declare no conflicts of interest.

References

1. Wilson, E.O. 1992. *The Diversity of Life*. Harvard University Press. Cambridge, MA.
2. Darwin, C. 1859 [2009]. *The Annotated Origin: A Facsimile of the First Edition of On the Origin of Species*. J. T. Costa, annotator. The Belknap Press of Harvard University Press. Cambridge, MA.
3. Brown, W.L. & E.O. Wilson. 1956. Character displacement. *Syst. Zool.* **5:** 49–64.
4. Grant, P.R. 1972. Convergent and divergent character displacement. *Biol. J. Linn. Soc.* **4:** 39–68.
5. Schluter, D. 2000. *The Ecology of Adaptive Radiation*. Oxford University Press. Oxford, UK.
6. Dayan, T. & D. Simberloff. 2005. Ecological and community-wide character displacement: the next generation. *Ecol. Lett.* **8:** 875–894.
7. Howard, D.J. 1993. Reinforcement: origin, dynamics, and fate of an evolutionary hypothesis. In *Hybrid Zones and the Evolutionary Process*. R. G. Harrison, Ed.: 46–69. Oxford University Press. New York.
8. Slatkin, M. 1980. Ecological character displacement. *Ecology* **61:** 163–177.
9. Schluter, D. 2001. Ecological character displacement. In *Evolutionary Ecology: Concepts and Case Studies*. C.W. Fox, D.A. Roff & D.J. Fairbairn, Eds.: 265–276. Oxford University Press. New York.
10. Crozier, R.H. 1974. Niche shape and genetical aspects of character displacement. *Am. Zool.* **14:** 1151–1157.
11. Odum, E.P. 1959. *Fundamentals of Ecology*. Saunders. Philadelphia, PA.
12. Dybzinski, R. & D. Tilman. 2009. Competition and coexistence in plant communities. In *The Princeton Guide to Ecology*. S.A. Levin, Ed.: 186–195. Princeton University Press. Princeton, NJ.
13. Pfennig, K.S. & D.W. Pfennig. 2009. Character displacement: ecological and reproductive responses to a common evolutionary problem. *Q. Rev. Biol.* **84:** 253–276.
14. Pfennig, D.W. & K.S. Pfennig. 2010. Character displacement and the origins of diversity. *Am. Nat.* **176:** S26–S44.
15. Endler, J.A. 1991. Interactions between predators and prey. In *Behavioural Ecology: An Evolutionary Approach*. J.R. Krebs & N.B. Davies, Eds.: 169–196. Blackwell. London.
16. Zuk, M. *et al.* 2006. Silent night: adaptive disappearance of a sexual signal in a parasitized population of field crickets. *Biol. Lett.* **2:** 521–524.
17. Gause, G.F. 1934. *The Struggle for Existence*. Williams and Wilkins. Baltimore.
18. Hardin, G. 1960. The competitive exclusion principle. *Science* **131:** 1292–1297.
19. Gröning, J. & A. Hochkirch. 2008. Reproductive interference between animal species. *Q. Rev. Biol.* **83:** 257–282.
20. Kishi, S. *et al.* 2009. Reproductive interference determines persistence and exclusion in species interactions. *J. Anim. Ecol.* **78:** 1043–1049.
21. Doebeli, M. 1996. An explicit genetic model for ecological character displacement. *Ecology* **77:** 510–520.
22. Taper, M.L. & T.J. Case. 1985. Quantitative genetic models for the coevolution of character displacement. *Ecology* **66:** 355–371.

23. Taper, M.L. & T.J. Case. 1992. Coevolution among competitors. *Oxford Surv. Evol. Biol.* **8:** 63–109.

24. Kirkpatrick, M. 2000. Reinforcement and divergence under assortative mating. *Proc. R. Soc. Lond. B* **267:** 1649–1655.

25. Liou, L.W. & T.D. Price. 1994. Speciation by reinforcement of premating isolation. *Evolution* **48:** 1451–1459.

26. Case, T.J. & M.L. Taper. 2000. Interspecific competition, environmental gradients, gene flow, and the coevolution of species' borders. *Am. Nat.* **155:** 583–605.

27. Schluter, D. & J.D. McPhail. 1992. Ecological character displacement and speciation in sticklebacks. *Am. Nat.* **140:** 85–108.

28. Losos, J.B. 2000. Ecological character displacement and the study of adaptation. *Proc. Natl. Acad.Sci. U.S.A.* **97:** 5693–5695.

29. Arthur, W. 1982. The evolutionary consequences of interspecific competition. *Adv. Ecol. Res.* **12:** 127–187.

30. Connell, J.H. 1980. Diversity and the coevolution of competitors, or the ghost of competition past. *Oikos* **35:** 131–138.

31. Servedio, M.R. *et al.* 2009. Reinforcement and learning. *Evol. Ecol.* **23:** 109–123.

32. Schlichting, C.D. & M. Pigliucci. 1998. *Phenotypic Evolution: A Reaction Norm Perspective.* Sinauer. Sunderland, MA.

33. Schlichting, C.D. 2008. Hidden reaction norms, cryptic genetic variation, and evolvability. *Ann. N. Y. Acad. Sci.* **1133:** 187–203.

34. Suzuki, Y. & H.F. Nijhout. 2006. Evolution of a polyphenism by genetic accommodation. *Science* **311:** 650–652.

35. Aubret, F. & R. Shine. 2009. Genetic assimilation and the postcolonization erosion of phenotypic plasticity in island Tiger Snakes. *Curr. Biol.* **19:** 1932–1936.

36. De Meester, L. 1996. Evolutionary potential and local genetic differentiation in a phenotypically plastic trait of a cyclical parthenogen, *Daphnia magna. Evolution* **50:** 1293–1298.

37. Pfennig, D.W. & P.J. Murphy. 2000. Character displacement in polyphenic tadpoles. *Evolution* **54:** 1738–1749.

38. Pfennig, D.W. & P.J. Murphy. 2002. How fluctuating competition and phenotypic plasticity mediate species divergence. *Evolution* **56:** 1217–1228.

39. Pfennig, D.W. *et al.* 2010. Phenotypic plasticity's impacts on diversification and speciation. *Trends Ecol. Evol.* **25:** 459–467.

40. Gilbert, S.F. & D. Epel. 2009. *Ecological Developmental Biology: Integrating Epigenetics, Medicine, and Evolution.* Sinauer Associates. Sunderland, MA.

41. Whitman, D.W. & A.A. Agrawal. 2009. What is phenotypic plasticity and why is it important? In *Phenotypic Plasticity of Insects.* D.W. Whitman & T.N. Ananthakrishnan, Eds.: 1–63. Science Publishers. Enfield, NH.

42. West-Eberhard, M.J. 2003. *Developmental Plasticity and Evolution.* Oxford University Press. New York.

43. Waddington, C.H. 1952. Selection of the genetic basis for an acquired character. *Nature* **169:** 278.

44. Waddington, C.H. 1953. Genetic assimilation of an acquired character. *Evolution* **7:** 118–126.

45. Rossiter, M.C. 1996. Incidence and consequences of inherited environmental effects. *Annu. Rev. Ecol. Syst.* **27:** 451–476.

46. Mousseau, T.A. & C.W. Fox. Eds. 1998. *Maternal Effects as Adaptations.* Oxford University Press. New York.

47. Payne, R.B. *et al.* 2000. Imprinting and the origin of parasite-host associations in brood-parasitic indigobirds, *Vidua chalybeata. Anim. Behav.* **59:** 69–81.

48. Engelstadter, J. & G.D.D. Hurst. 2009. The ecology and evolution of microbes that manipulate host reproduction. *Annu. Rev. Ecol. Evol. Syst.* **40:** 127–149.

49. Emlen, D.J. *et al.* 2007. On the origin and evolutionary diversification of beetle horns. *Proc. Natl. Acad. Sci. U.S.A.* **104:** 8661–8668.

50. Ledón-Rettig, C.C. *et al.* 2008. Ancestral variation and the potential for genetic accommodation in larval amphibians: implications for the evolution of novel feeding strategies. *Evol. Dev.* **10:** 316–325.

51. Matsuda, R. 1987. *Animal Evolution in Changing Environments with Special Reference to Abnormal Metamorphosis.* John Wiley and Sons. New York.

52. Hall, B.K. 1999. *Evolutionary Developmental Biology.* Kluwer. Dordrecht.

53. Moczek, A.P. *et al.* 2011. The role of developmental plasticity in evolutionary innovation. *Proc. R. Soc. Lond. B* **278:** 2705–2713.

54. Felsenstein, J. 1981. Skepticism toward Santa Rosalia, or why are there so few kinds of animals? *Evolution* **35:** 124–138.

55. Servedio, M.R. & M.A.F. Noor. 2003. The role of reinforcement in speciation: theory and data. *Annu. Rev. Ecol. Syst.* **34:** 339–364.

56. Coyne, J.A. & H.A. Orr. 2004. *Speciation.* Sinauer. Sunderland, MA.

57. Ortiz-Barrientos, D. & M.A.F. Noor. 2005. Evidence for a one-allele assortative mating locus. *Science* **310:** 1467–1467.

58. Noor, M.A. 1995. Speciation driven by natural selection in *Drosophila. Nature* **375:** 674–675.

59. Ortiz-Barrientos, D. *et al.* 2004. The genetics of speciation by reinforcement. *PLos Biol.* **2:** 2256–2263.

60. Barnwell, C.V. & M.A.F. Noor. 2008. Failure to replicate two mate preference QTLs across multiple strains of *Drosophila pseudoobscura. J. Hered.* **99:** 653–656.

61. Brower, A.V.Z. 1996. Parallel race formation and the evolution of mimicry in *Heliconius* butterflies: a phylogenenetic hypothesis from mitochondrial DNA sequences. *Evolution* **50:** 195–221.

62. Turner, J.R.G. 1981. Adaptation and evolution in *Heliconius*—a defense of neodarwinism. *Annu. Rev. Ecol. Syst.* **12:** 99–121.

63. Mallet, J. & L.E. Gilbert. 1995. Why are there so many mimicry rings—correlations between habitat, behavior and mimicry in *Heliconius* butterflies. *Biol. J. Linn. Soc.* **55:** 159–180.

64. Kronforst, M.R. *et al.* 2006. Linkage of butterfly mate preference and wing color preference cue at the genomic location of wingless. *Proc. Natl. Acad. Sci. U.S.A.* **103:** 6575–6580.

65. Estrada, C. & C.D. Jiggins. 2008. Interspecific sexual attraction because of convergence in warning colouration:

is there a conflict between natural and sexual selection in mimetic species? *J. Evol. Biol.* **21:** 749–760.

66. Chamberlain, N.L. *et al.* 2009. Polymorphic butterfly reveals the missing link in ecological speciation. *Science* **326:** 847–850.

67. Saetre, G.P. *et al.* 1997. A sexually selected character displacement in flycatchers reinforces premating isolation. *Nature* **387:** 589–592.

68. Backstrom, N. *et al.* 2010. A high-density scan of the Z chromosome in *Ficedula* flycatchers reveals candidate loci for diversifying selection. *Evolution* **64:** 3461–3475.

69. McPhail, J.D. 1984. Ecology and evolution of sympatric sticklebacks (*Gasterosteus*): morphological and genetic evidence for a species pair in Enos Lake, British Columbia. *Can. J. Zool.-Rev. Can. Zool.* **62:** 1402–1408.

70. McPhail, J.D. 1992. Ecology and evolution of sympatric sticklebacks (*Gasterosteus*): evidence for a species pair in Paxton Lake, Texada Island, British Columbia. *Can. J. Zool.-Rev. Can. Zool.* **70:** 361–369.

71. Hatfield, T. 1997. Genetic divergence in adaptive characters between sympatric species of sticklebacks. *Am. Nat.* **149:** 1009–1029.

72. Hatfield, T. & D. Schluter. 1999. Ecological speciation in sticklebacks: environment-dependent hybrid fitness. *Evolution* **53:** 866–873.

73. Smith, R.A. & M.D. Rausher. 2008. Experimental evidence that selection favors character displacement in the ivyleaf morning glory. *Am. Nat.* **171:** 1–9.

74. Martin, R.A. & D.W. Pfennig. 2011. Evaluating the targets of selection during character displacement. *Evolution* **65:** 2946–2958.

75. Butlin, R.K. 2005. Recombination and speciation. *Mol. Ecol.* **14:** 2621–2635.

76. Hoffmann, A.A. & L.H. Rieseberg. 2008. Revisiting the impact of inversions in evolution: from population genetic markers to drivers of adaptive shifts and speciation? *Annu. Rev. Ecol. Evol. Syst.* **39:** 21–42.

77. Behm, J.E. *et al.* 2010. Breakdown in postmating isolation and the collapse of a species pair through hybridization. *Am. Nat.* **175:** 11–26.

78. Pfennig, K.S. 2003. A test of alternative hypotheses for the evolution of reproductive isolation between spadefoot toads: support for the reinforcement hypothesis. *Evolution* **57:** 2842–2851.

79. Grant, B.R. & P.R. Grant. 2008. Fission and fusion of Darwin's finches populations. *Philos. Trans. R. Soc. B Biol. Sci.* **363:** 2821–2829.

80. Reifova, R. *et al.* 2011. Ecological character displacement in the face of gene flow: evidence from two species of nightingales. *BMC Evol. Biol.* **11:** 138.

81. Noor, M.A.F. *et al.* 2001. Chromosomal inversions and the reproductive isolation of species. *Proc. Natl. Acad. Sci. U.S.A.* **98:** 12084–12088.

82. Noor, M.A.F. & S.M. Bennett. 2009. Islands of speciation or mirages in the desert? Examining the role of restricted recombination in maintaining species. *Heredity* **103:** 439–444.

83. King, M.C. & A.C. Wilson. 1975. Evolution at two levels in humans and chimpanzees. *Science* **188:** 107–116.

84. Raff, R.A. & T.C. Kauffman. 1983. *Genes, Embryos, and Evolution.* MacMillan. New York.

85. Carroll, S.B. *et al.* 2001. *From DNA to Diversity: Molecular Genetics and the Evolution of Animal Design.* Blackwell. Malden, MA.

86. Wilkins, A.S. 2002. *The Evolution of Developmental Pathways.* Sinauer Associates. Sunderland, MA.

87. Wray, G.A. 2007. The evolutionary significance of cis-regulatory mutations. *Nat. Rev. Genet.* **8:** 206–216.

88. Wittkopp, P.J. & G. Kalay. 2012. *Cis*-regulatory elements: molecular mechanisms and evolutionary processes underlying divergence. *Nature Reviews Genetics* **13:** 59–69.

89. Grant, P.R. & B.R. Grant. 2008. *How and Why Species Multiply: The Radiation of Darwin's Finches.* Princeton University Press. Princeton, NJ.

90. Lack, D. 1947. *Darwin's Finches.* Cambridge University Press. Cambridge.

91. Podos, J. 2001. Correlated evolution of morphology and vocal signal structure in Darwin's finches. *Nature* **409:** 185–188.

92. Abzhanov, A. *et al.* 2004. Bmp4 and morphological variation of beaks in Darwin's finches. *Science* **305:** 1462–1465.

93. Abzhanov, A. *et al.* 2006. The calmodulin pathway and evolution of elongate beak morphology in Darwin's finches. *Nature* **442:** 563–567.

94. McGraw, E.A. *et al.* 2011. High-dimensional variance partitioning reveals the modular genetic basis of adaptive divergence in gene expression during reproductive character displacement. *Evolution* **65:** 3126–3137.

95. Hopkins, R. & M.D. Rausher. 2011. Identification of two genes causing reinforcement in the Texas wildflower *Phlox drummondii.* *Nature* **469:** 411–414.

96. Hoekstra, H.E. & J.A. Coyne. 2007. The locus of evolution: evo devo and the genetics of adaptation. *Evolution* **61:** 995–1016.

97. Razeto-Barry, P. & K. Maldonado. 2011. Adaptive *cis*-regulatory changes may involve few mutations. *Evolution* **65:** 3332–3335.

98. Pavey, S.A. *et al.* 2010. The role of gene expression in ecological speciation. *Ann. N. Y. Acad. Sci.* **1206:** 110–129.

99. Pfennig, D.W. *et al.* 2006. Ecological opportunity and phenotypic plasticity interact to promote character displacement and species coexistence. *Ecology* **87:** 769–779.

100. Schluter, D. 2002. Character displacement. In *Encyclopedia of Evolution.* M. Pagel, Ed.: 149–150. Oxford University Press. Oxford, UK.

101. Pfennig, D.W. & M. McGee. 2010. Resource polyphenism increases species richness: a test of the hypothesis. *Phil. Trans. R. Soc. Lond B* **365:** 577–591.

102. Chevin, L.M. *et al.* 2010. Adaptation, plasticity, and extinction in a changing environment: towards a predictive theory. *PLos Biol.* **8:** e1000357.

103. Schoener, T.W. 1974. Resource partitioning in ecological communities. *Science* **185:** 27–39.

104. Case, T.J. & R. Sidell. 1983. Pattern and chance in the structure of model and natural communities. *Evolution* **37:** 832–849.

105. Losos, J.B. 1990. A phylogenetic analysis of character displacement. *Evolution* **44**: 558–569.

106. Agrawal, A.A. 2001. Phenotypic plasticity in the interactions and evolution of species. *Science* **294**: 321–326.

107. Fordyce, J.A. 2006. The evolutionary consequences of ecological interactions mediated through phenotypic plasticity. *J. Exp. Biol.* **209**: 2377–2383.

108. West-Eberhard, M.J. 1992. Behavior and evolution. In *Molds, Molecules, and Metazoa: Growing Points in Evolutionary Biology*. P.R. Grant & H.S. Horn, Eds.: 57–75. Princeton University Press. Princeton, NJ.

109. Robinson, B.W. & D.S. Wilson. 1994. Character release and displacement in fish: a neglected literature. *Am. Nat.* **144**: 596–627.

110. Nobel, P.S. 1997. Root distribution and seasonal production in the northwestern Sonoran desert for a C3 subshrub, a C4 bunchgrass, and a CAM leaf succulent. *Am. J. Bot.* **84**: 949–955.

111. Cahill, J.F. *et al.* 2010. Plants integrate information about nutrients and neighbors. *Science* **328**: 1657.

112. Mahall, B.E. & R.M. Callaway. 1991. Root communication among desert shrubs. *Proc. Natl. Acad. Sci.* **88**: 874–876.

113. Gersani, M. *et al.* 2001. Tragedy of the commons as a result of root competition. *J. Ecol.* **89**: 660–669.

114. Schmitt, J. *et al.* 1999. Manipulative approaches to testing adaptive plasticity: phytochrome-mediated shade-avoidance responses in plants. *Am. Nat.* **154**: S43–S54.

115. Rhee, G.-Y. 1978. Effects of N : P atomic ratios and nitrate limitation on algal growth, cell composition, and nitrate uptake. *Limnol. Oceanogr.* **23**: 10–25.

116. Pfennig, K.S. 2007. Facultative mate choice drives adaptive hybridization. *Science* **318**: 965–967.

117. Pfennig, D.W. & R.A. Martin. 2009. A maternal effect mediates rapid population divergence and character displacement in spadefoot toads. *Evolution* **63**: 898–909.

118. Pfennig, D.W. & R.A. Martin. 2010. Evolution of character displacement in spadefoot toads: different proximate mechanisms in different species. *Evolution* **64**: 2331–2341.

119. Donohue, K. & J. Schmitt. 1998. Maternal environmental effects in plants: adaptive plasticity? In *Maternal Effects as Adaptations*. T.A. Mousseau & C.W. Fox, Eds.: 137–158. Oxford University Press. New York.

120. Allen, R.M. *et al.* 2008. Offspring size plasticity in response to intraspecific competition: an adaptive maternal effect across life-history stages. *Am. Nat.* **171**: 225–237.

121. Badyaev, A.V. *et al.* 2002. Sex-biased hatching order and adaptive population divergence in a passerine bird. *Science* **295**: 316–318.

122. Badyaev, A.V. 2008. Maternal effects as generators of evolutionary change: a reassessment. *Ann. N. Y. Acad. Sci.* **1133**: 151–161.

123. Räsänen, K. & L.E.B. Kruuk. 2007. Maternal effects and evolution at ecological time-scales. *Funct. Ecol.* **21**: 408–421.

124. Galloway, L.F. & J.R. Etterson. 2007. Transgenerational plasticity is adaptive in the wild. *Science* **318**: 1134–1136.

125. Uller, T. 2008. Developmental plasticity and the evolution of parental effects. *Trends Ecol. Evol.* **23**: 432–438.

126. Plaistow, S.J. *et al.* 2006. Context-dependent intergenerational effects: the interaction between past and present environments and its effect on population dynamics. *Am. Nat.* **167**: 206–215.

127. Grant, B.R. & P.R. Grant. 2010. Songs of Darwin's finches diverge when a new species enters the community. *Proc. Natl. Acad. Sci. U.S.A.* **107**: 20156–20163.

128. Kozak, G.M. *et al.* 2011. Sexual imprinting on ecologically divergent traits leads to sexual isolation in sticklebacks. *Proc. R. Soc. B* **278**: 2604–2810.

129. Papaj, D.R. & R.J. Prokopy. 1989. Ecological and evolutionary aspects of learning in phytophagous insects. *Annu. Rev. Entomol.* **34**: 315–350.

130. Price, T. 2008. *Speciation in Birds*. Roberts and Company Publishers. Greenwood Village, CO.

131. Milligan, B.G. 1985. Evolutionary divergence and character displacement in two phenotypically-variable competing species. *Evolution* **39**: 1207–1222.

132. Rice, A.M. & D.W. Pfennig. 2007. Character displacement: *in situ* evolution of novel phenotypes or sorting of pre-existing variation? *J. Evol. Biol.* **20**: 448–459.

133. Grant, P.R. & B.R. Grant. 2006. Evolution of character displacement in Darwin's finches. *Science* **313**: 224–226.

134. Barrett, R.D.H. & D. Schluter. 2008. Adaptation from standing genetic variation. *Trends Ecol. Evol.* **23**: 38–44.

135. Phillips, P.C. 1996. Waiting for a compensatory mutation: phase zero of the shifting-balance process. *Genet. Res.* **67**: 271–283.

136. Gibson, G. & I. Dworkin. 2004. Uncovering cryptic genetic variation. *Nat. Rev. Genet.* **5**: 1199–1212.

137. Ledón-Rettig, C.C. *et al.* 2010. Diet and hormone manipulations reveal cryptic genetic variation: implications for the evolution of novel feeding strategies. *Proc. R. Soc. Lond. B* **277**: 3569–3578.

138. West-Eberhard, M.J. 1989. Phenotypic plasticity and the origins of diversity. *Annu. Rev. Ecol. Syst.* **20**: 249–278.

139. West-Eberhard, M.J. 2005. Developmental plasticity and the origin of species differences. *Proc. Natl. Acad. Sci. U.S.A.* **102**: 6543–6549.

140. Smith, T.B. & S. Skúlason. 1996. Evolutionary significance of resource polymorphisms in fishes, amphibians, and birds. *Annu. Rev. Ecol. Syst.* **27**: 111–133.

141. Emlen, D.J. 2008. The evolution of animal weapons. *Annu. Rev. Ecol. Syst. Evol.* **39**: 387–413.

142. Gross, M.R. 1996. Alternative reproductive strategies and tactics: diversity within sexes. *Trends Ecol. Evol.* **11**: 92–98.

143. Levene, H. 1953. Genetic equilibrium when more than one ecological niche is available. *Am. Nat.* **87**: 331–333.

144. Clarke, B.C. 1975. The contribution of ecological genetics to evolutionary theory: detecting the direct effects of natural selection on particularly polymorphic loci. *Genetics* **79**: 101–113.

145. Schluter, D. 1996. Adaptive radiation along genetic lines of least resistance. *Evolution* **50**: 1766–1774.

146. Baldwin, J.M. 1896. A new factor in evolution. *Am. Nat.* **30**: 441–451.

147. Schmalhausen, I.I. 1949 [1986]. *Factors of Evolution: The Theory of Stabilizing Selection.* University of Chicago Press. Chicago, IL.

148. Pigliucci, M. & C.J. Murren. 2003. Genetic assimilation and a possible evolutionary paradox: can macroevolution sometimes be so fast as to pass us by? *Evolution* **57:** 1455–1464.

149. Pál, C. & I. Miklos. 1999. Epigenetic inheritance, genetic assimilation and speciation. *J. Theor. Biol.* **200:** 19–37.

150. Price, T.D. *et al.* 2003. The role of phenotypic plasticity in driving genetic evolution. *Proc. R. Soc. Lond. B* **270:** 1433–1440.

151. Moczek, A.P. 2008. On the origins of novelty in development and evolution. *Bioessays* **30:** 432–447.

152. Schwander, T. & O. Leimar. 2011. Genes as leaders and followers in evolution. *Trends Ecol. Evol.* **26:** 143–151.

153. Lande, R. 2009. Adaptation to an extraordinary environment by evolution of phenotypic plasticity and genetic assimilation. *J. Evol. Biol.* **22:** 1435–1446.

154. Masel, J. *et al.* 2007. The loss of adaptive plasticity during long periods of environmental stasis. *Am. Nat.* **169:** 38–46.

155. Lahti, D.C. *et al.* 2009. Relaxed selection in the wild. *Trends Ecol. Evol.* **24:** 487–496.

156. Schlichting, C.D. 2004. The role of phenotypic plasticity in diversification. In *Phenotypic Plasticity: Functional and Conceptual Approaches.* T.J. DeWitt & S.M. Scheiner, Eds.: 191–200. Oxford University Press. New York.

157. Pijanowska, J. *et al.* 2007. Phenotypic plasticity within *Daphnia longispina* complex: differences between parental and hybrid clones. *Pol. J. Ecol.* **55:** 761–769.

158. Relyea, R.A. 2002. Costs of phenotypic plasticity. *Am. Nat.* **159:** 272–282.

159. Auld, J.R. *et al.* 2010. Re-evaluating the costs and limits of adaptive phenotypic plasticity. *Proc. R. Soc. Lond. B.* **277:** 503–511.

160. Fenchel, T. 1975. Character displacement and coexistence in mud snails (Hydrobiidae). *Oecologia* **20:** 19–32.

161. Diamond, J. *et al.* 1989. Rapid evolution of character displacement in Myzomelid honeyeaters. *Am. Nat.* **134:** 675–708.

162. Yom-Tov, Y. *et al.* 1999. Competition, coexistence, and adaptation amongst rodent invaders to Pacific and New Zealand islands. *J. Biogeogr.* **26:** 947–958.

163. Pfennig, D.W. & P.J. Murphy. 2003. A test of alternative hypotheses for character divergence between coexisting species. *Ecology* **84:** 1288–1297.

164. Hoskin, C.J. *et al.* 2005. Reinforcement drives rapid allopatric speciation. *Nature* **437:** 1353–1356.

165. Hansen, T.F. *et al.* 2000. Comparative analysis of character displacement and spatial adaptations as illustrated by the evolution of *Dalechampia* blossoms. *Am. Nat.* **156:** S17–S34.

166. Schluter, D. & L.M. Nagel. 1995. Parallel speciation by natural selection. *Am. Nat.* **146:** 292–301.

167. Marko, P.B. 2005. An intraspecific comparative analysis of character divergence between sympatric species. *Evolution* **59:** 554–564.

168. Matocq, M.D. & P.J. Murphy. 2007. Fine-scale phenotypic change across a species transition zone in the genus *Neotoma*: disentangling independent evolution from phylogenetic history. *Evolution* **61:** 2544–2557.

169. Rice, A.M. *et al.* 2009. Parallel evolution and ecological selection: replicated character displacement in spadefoot toads. *Proc. R. Soc. Lond. B* **276:** 4189–4196.

170. Adams, D.C. 2010. Parallel evolution of character displacement driven by competitive selection in terrestrial salamanders. *BMC Evol. Biol.* **10:** 1–10.

171. Rundle, H.D. & D. Schluter. 2004. Natural selection and ecological speciation in sticklebacks. In *Adaptive Speciation.* U. Dieckmann, M. Doebeli, J.A.J. Metz & D. Tautz, Eds.: 192–209. Cambirdge University Press. Cambridge, UK.

172. Wund, M.A. *et al.* 2008. A test of the "flexible stem" model of evolution: ancestral plasticity, genetic accommodation, and morphological divergence in the threespine stickleback radiation. *Am. Nat.* **172:** 449–462.

173. Skúlason, S. *et al.* 1999. Sympatric morphs, populations and speciation in freshwater fish with emphasis on arctic charr. In *Evolution of Biological Diversity.* A.E. Magurran & R.M. May, Eds.: 70–92. Oxford University Press. Oxford, UK.

174. Meyer, A. 1987. Phenotypic plasticity and heterochrony in *Ciclasoma managuense* (Pisces, Cichlidae) and their implications for speciation in cichlid fishes. *Evolution* **41:** 1357–1369.

175. Robinson, B.W. & D.S. Wilson. 1996. Genetic variation and phenotypic plasticity in a trophically polymorphic population of pumpkinseed sunfish (*Lepomis gibbosus*). *Evol. Ecol.* **10:** 631–652.

176. Pfennig, D.W. *et al.* 2007. Field and experimental evidence for competition's role in phenotypic divergence. *Evolution* **61:** 257–271.

177. Rice, A.M. & D.W. Pfennig. 2008. Analysis of range expansion in two species undergoing character displacement: why might invaders generally "win" during character displacement? *J. Evol. Biol.* **21:** 696–704.

178. Fox, C.W. & T.A. Mousseau. 1998. Maternal effects as adaptations for transgenerational phenotypic plasticity in insects. In *Maternal Effects as Adaptations.* T.A. Mousseau & C.W. Fox, Eds.: 159–177. Oxford University Press. New York.

179. Losos, J.B. 2009. *Lizards in an Evolutionary Tree: Ecology and Adaptive Radiation of Anoles.* University of California Press. Berkeley, CA.

180. Williams, E.E. 1972. The origin of faunas. Evolution of lizard congeners in a complex island fauna: a trial analysis. *Evol. Biol.* **6:** 47–89.

181. Losos, J.B. *et al.* 2000. Evolutionary implications of phenotypic plasticity in the hindlimb of the lizard *Anolis sagrei*. *Evolution* **54:** 301–305.

182. Schoener, T.W. 1970. Size patterns in West Indian *Anolis* lizards. 2. Correlations with the size of particular sympatric species—-displacement and convergence. *Am. Nat.* **104:** 155–174.

183. Tyerman, J.G. *et al.* 2008. Experimental demonstration of ecological character displacement. *BMC Evol. Biol.* **8:** 34.

184. Kassen, R. 2009. Toward a general theory of adaptive radiation: insights from microbial experimental evolution. *Ann. N.Y. Acad. Sci.* **1168:** 3–22.

185. Pfennig, D.W. & K.S. Pfennig. forthcoming. *Evolution's Wedge: Competition and the Origins of Diversity.* University of California Press. Berkeley, CA.